BEYOND THE CRIME LAB

BEYOND THE CRIME LAB

The New Science of Investigation

Revised Edition

JON ZONDERMAN

John Wiley & Sons, Inc.

New York • Chichester • Weinheim • Brisbane • Singapore • Toronto

This book is printed on acid-free paper. ∞

Copyright © 1999 by Jon Zonderman. All rights reserved.
Published by John Wiley & Sons, Inc.
Published simultaneously in Canada.
Design and production by Navta Associates, Inc.

First edition copyright © 1990 by John Wiley & Sons, Inc.

This publication is designed to provide accurate and authoritative information in regard to the subject matter covered. It is sold with the understanding that the publisher is not engaged in rendering professional services. If professional advice or other expert assistance is required, the services of a competent professional should be sought.

Library of Congress Cataloging-in-Publication Data:
Zonderman, Jon.
    Beyond the crime lab : the new science of investigation / Jon
Zonderman. — Rev. ed.
        p.   cm.
    Includes index.
    ISBN 0-471-25466-5 (cloth : alk. paper)
    1. Criminal investigation. 2. Criminals—Identification.
3. Crime laboratories. 4. Electronics in criminal investigation.
I. Title
HV8073.Z66 1998
363.25—dc21   98-14283

Printed in the United States of America

10 9 8 7 6 5 4 3 2 1

For Anna, Jacob, & Laurel

Contents

Foreword

Forensic science has emerged as a significant element in efforts to solve crime while maintaining a high quality of justice. The value of physical evidence has been demonstrated in all aspects of criminal investigation, and law-enforcement officials have become increasingly dependent on laboratory results for evidence not obtainable by other means. As science and technology continue to advance, the importance and value of physical evidence in criminal investigation also will continue to grow.

Virtually any type of material can become physical evidence. It can be as small as a dust particle or as large as an airplane. It can be in the form of a gas or a liquid. It may consist of only one simple pattern, or it may contain thousands of pages of documents and photographs.

There are hundreds of methods used in the forensic laboratory. Many of these methods are common biological, immunological, biochemical, microscopic, chemical, physical, or instrumental techniques used in other clinics or in research or scientific laboratories. Other methods are unique to the forensic field. Forensic analysis is concerned not only with the recognition and identification of unknown substances, but also with the individualization and reconstruction of a variety of evidence, events, or conditions.

The aim of examining physical evidence is to provide useful information and scientific truth to solve cases and protect the innocent and society. However, decisions about the extent to which physical evidence will be used in criminal investigations usually are not made by forensic scientists. In the crime-scene search and initial investigative stages, these decisions usually are made by police officers, criminal investigators, or evidence technicians. In the adjudicative stages, physical evidence usually is used by prosecutors and defense attorneys. There is no guarantee that either the investigators or the attorneys will understand the potential of physical evidence well enough to make the proper decisions.

Along with progress in the forensic-science discipline, we also have witnessed the proliferation of various publications. These books and articles generally fall into two groups: (1) very technically oriented publications, and (2) semischolarly, basic materials. The former type is aimed at forensic professionals; the latter is for leisure reading. There is a gap between these two types of publications and a need to relate the progress in forensic science to police officers, attorneys, and the general public.

Jon Zonderman has devoted an enormous amount of time and energy to research and develop materials related to the advances in forensic science. This book bridges the gap between the basic and the advanced materials. As more police officers receive special training in forensic sciences, and as attorneys and the general public acquire updated information, better use will be made of physical evidence and forensic science.

Henry C. Lee, Ph.D.
Commissioner of Public Safety, State of Connecticut
Chief Criminalist/Director, Connecticut State Police
Forensic Science Laboratory, Retired
Professor of Forensic Science, University of New Haven

Preface to the Revised Edition

When I was researching and writing *Beyond the Crime Lab,* in 1988 and 1989, forensic science was still not a part of American popular culture. The "true life" crime story still took a back seat to more sedate murder mysteries in literature, and the courtroom drama and shoot-'em-up aspects of police television were still the strong points of any script's plotting.

Today, as this newly updated and revised edition is readied for print, Americans are riveted to their television sets by the *trial du jour,* be it O. J. Simpson, Timothy McVeigh, the Boston Nanny, or the Unabomber—and people actually argue about the forensic evidence presented in the case. Was O. J.'s bloody glove planted? How can Kaczynski pretend not to be the Unabomber when an explosive device with all of the relevant "signatures" was found in his cabin?

Police shows focus on the forensics of every case that needs to be solved, and the scriptwriters don't just give us the perfunctory detective saying, "Ballistics says it's a match." Instead, they get a little education in there at the same time: "The slug in the wall was too damaged to get any information about lands and grooves, only enough to say it was the same caliber as the suspect's gun." Medical examiners are prominent characters, and legal arguments often hinge on forensic minutia.

Cable television chimes in with a weekly show about forensics on the Discovery channel, and some of the country's most prominent forensic scientists have nearly become folk heroes for their daily appearances on *Good Morning America, CNN News,* and legal shows.

The independent counsel investigating the president has even turned to forensic investigations, with handwriting experts and paper-and-ink specialists poring over the Rose Law Firm billing records found in the White House to see whether any of them had been tampered with; and more recently subjecting Monica Lewinsky's clothing to testing for President Clinton's bodily fluids.

Forensic science is increasingly coming to be a widely taught academic discipline, as well. On the college level, students can take specialized courses in chemistry, biology, and physical science curriculums, and they can pursue forensic-science programs through a few specialized schools. At the high school level, in science magnet schools and as part of twenty-first-century school-to-work programs, the job of forensic laboratory technician is coming to be seen as a legitimate field of study.

Still, if you ask a random sampling of 12 jury-eligible Americans to explain how DNA typing works, or even how blood typing works, few if any could reasonably describe the technique. They wouldn't be much more helpful about conventional fingerprinting, or ballistics, or firearms and tool marks, or chemical analysis of bomb residue.

We are fascinated by what repulses us—skeletal remains and serial rapists. They are our worst nightmares, yet we flock to the bookshelves on which they are described in million-copy detail by the likes of Patricia Cornwell and Thomas Harris.

This book is designed to fill in the knowledge gaps, giving a nontechnical reader enough insight and a baseline of information to understand in broad terms the arguments being made by the experts and the ever-more-scientifically complex story lines. It is also meant to give readers enough understanding of the issues to allow them to ask whether science is, on balance, more helpful in proving guilt than it is harmful in impinging on our civil liberties.

As we enter the new millennium, we all need to be mindful of ceding our political and civil freedoms to expertise of any kind. Do we really want to live in a society where we are constantly under the surveillance of the science police?

Jon Zonderman
Orange, Connecticut
1998

Preface to the First Edition

Each year, U.S. police investigate more than 20,000 killings and millions of other crimes. Increasingly, they turn to the forensic scientists and criminologists who work in crime laboratories and medical examiner's offices—and to a host of consulting specialists from universities and private forensics labs—to provide them with scientific information about physical evidence that will link victim to evidence to suspect in a way that judges and juries will find credible. They are also turning more often to the masses of information in computerized databases to help them draw conclusions about victims and suspects. Finally, they are turning to an array of sophisticated, high-technology surveillance equipment to help them keep close watch on individuals who are under investigation.

This book is about those techniques, how they are used, and how they might be abused. It is about dedicated people who work to master the myriad, complicated techniques of scientific inquiry. It is about the political and bureaucratic processes, as well as the drastic increases in demand for these people's services, that conspire to make their work difficult. It is about too much work, too little time and training, and a judicial system so overburdened that the professionals who do scientific sleuthing feel ground down. It is about basic scientific research and how that research is transformed into practical use in a legal context. It is about how increased scientific information and technology could be used to create a safer society, and about how that information and technology could be abused to create a less free society.

Jon Zonderman
Orange, Connecticut
1990

INTRODUCTION

It probably is a happy coincidence that what we today would consider to be modern science and political philosophy have developed somewhat in parallel since the late 1700s. In fact, some would argue that the ideas of democracy and capitalism are necessary for the full flourishing of scientific thought and expertise.

The strains between social organization and individual rights in Western societies—the idea that there is a civil leadership instead of (or as well as) a religious or royal one, and that, in many instances, the civil authorities take precedence—led to the notion that crimes are committed not only against an individual, against the church, or against the king, but also against "the people," "the state," or "the Commonwealth."

Since at least the eighteenth century, the civil societies of the West have organized the fight against crime and have placed people in positions of authority with responsibility for upholding the law. No longer was the sheriff merely an instrument of royalty, set on the peasantry to inflict the royal whim, but a person whose duty it was to see that objective laws were obeyed by all.

At about the same time, scientists and natural philosophers, using ever more sophisticated instruments in their observations, were rapidly incorporating new knowledge of the world around them and beginning to disseminate that knowledge to a wider audience than ever before.

By the time of the Industrial Revolution, cities were beginning to hire police forces to uphold the laws of the civil community. In 1810 the French *Sureté* ("Security") was created to try to stop a wave of crime in Paris. The founder, Eugene François Vidocq, was himself a former convict, and he brought on as his trusted subordinates other former convicts. Unfortunately, many of these men had a difficult time breaking their criminal habits and, when they weren't policing, were an active part of the French underworld.

The English concept of civil liberties made Britain suspicious of any kind of police force, and it was not until 1829 that London became so crime infested that Robert Peel, the Home Secretary, went against public opinion and led the fight in the House of Commons to create a city police force. This new force replaced the Bow Street Runners, a group of about a dozen investigators organized in the mid-eighteenth century by Henry Fielding, then justice of the peace. Many of the runners were former "thief takers," private bounty hunters who caught criminals in exchange for rewards. In fact, any citizen could hire a Bow Street Runner when he wasn't on duty for the justice of the peace.

One thousand men were hired for the new London police force that replaced the runners; they wore blue tailcoats, gray trousers, and top hats, to give the impression that they were civilians rather than military. In 1842, 12 such police officers were removed from the uniformed force to create the city's first detective bureau. They worked out of three small rooms in Scotland Yard, and, despite its growth in size and change of location, the current London detective force still is known by that name. Charles Dickens made the force famous through the character Inspector Bucket of Scotland Yard, in his novel *Bleak House.*

By the 1860s, the notion of a detective investigator was well entrenched in the public perception, and these detectives were turning to the analysis of physical evidence—the natural or synthetic substances that appeared at crime scenes—in addition to questioning crime witnesses to get information about who may have committed the crime.

The fields of forensic science and criminalistics began to coalesce around this time. Chemical tests to determine whether a substance is blood, and what species of animal the blood may have come from, date to the 1860s. Working in various parts of the world, scientists and police detectives began the long process of determining that fingerprints are individual, that the prints left at the scene of a crime can be matched with prints taken from criminal suspects, under optimum conditions, and that permanent records of these prints would be highly valuable in keeping track of criminals.

In the 1880s, Sir Arthur Conan Doyle, a trained physician, began writing his Sherlock Holmes detective stories. To Holmes, everything was "elementary," a wonderful pun when one considers that Holmes's crime solving often was as firmly based on chemistry as it was on inductive logic. The Holmes stories, it is said, inspired Edmond Locard to set up the world's first forensic-science laboratory in France in 1910. The word *forensic* means suitable for the law court, and it was this organized approach to the analysis of physical evidence for legal presentation that gave rise to the field.

FORENSIC SCIENCE IN THE UNITED STATES

In the United States, scientific analysis of physical evidence was a little slower to catch on. Throughout the early years of the twentieth century, police forces occasionally turned to scientists to help analyze evidence. These scholars, however, had neither the proper facilities to conduct systematic investigations that would stand up to the rigors of the legal system nor the training to present their findings to judges and juries.

In 1930, the Federal Bureau of Investigation's first director, J. Edgar Hoover, decided to make the FBI the font of scientific criminal investigative knowledge in the United States. He had his special agents throughout the country contact experts in a number of scientific fields and ask them for their thoughts on how to build and staff a forensic-science laboratory and how to train forensic scientists. The resulting laboratory opened on November 24, 1932. It handled 20 cases in its first week. Today, it does thousands of examinations each week, both for its own agents working on federal cases and, on a consulting basis, for public-safety agencies throughout the United States and its protectorates.

The first home of the FBI's "Criminological Laboratory" was in Room 802 of the Old Southern Railway Building at 13th Street and Pennsylvania Avenue NW, in Washington, D.C. By June 1933, the name was changed to the "Technical Laboratory" and, in September 1935, the laboratory moved to larger quarters in the Department of Justice building at 9th Street and Pennsylvania Avenue NW. In 1943, the name was changed again, this time to the "FBI Laboratory," the name it still bears. The lab was to make one more move, in September 1975, to its present site in the J. Edgar Hoover FBI Building at 10th Street and Pennsylvania Avenue NW.

The FBI Lab has become the country's leader in many areas of forensic science and continues to conduct extensive training for forensic-science laboratory personnel throughout the country in both technical and administrative areas.

The FBI Lab's reputation has fallen on hard times through the 1990s. A 1996 investigation of the lab by the Justice Department's Inspector General found not only shoddy science, but also bureaucratic infighting and lack of leadership among the lab's top officials. In 1997, FBI Director Louis Freeh appointed the first outsider, a long-time administrator of government research labs, to run the FBI Lab.

ADVANCES IN FORENSIC SCIENCE

Although scientists made great strides in criminal investigation before 1960, since then, there have been quantum leaps in forensic science and, especially, technology. Today's scientific criminal investigator can gather and analyze far more information about the physical—and even behavioral—evidence that appears at a crime scene than their counterparts in the 1960s dreamed possible.

Police detectives and prosecutors have gained powerful new tools in their fight against crime through scientific breakthroughs in the study of DNA; advances in instrumentation such as microscopy, chromatography, and mass spectroscopy; the ability to lift latent fingerprints from almost any kind of organic or inorganic surface using such innovative methods as laser beams and the vapors from household adhesives such as Superglue; and the revolution in computers, communications, and electronics. Advances in both behavioral and biochemical sciences suggest that in the future, (a) we may understand both the physiological and the environmental causes of crime, (b) there may be medical and scientific ways to correct these causes, and (c) neither punishment nor do-gooder attempts to "change" criminals may be possible.

In the future, every cop, not just the police and scientists of the forensic laboratory, will be a science cop. A concurrent trend has been toward the increased use of forensic scientific analysis in civil litigation; forensic experts regularly appear in cases of product liability, fraud, air and vehicle accidents, and family matters such as spouse and child abuse. With this trend has come a booming business of scientific expert witnesses. In the future, not only will every cop be a science cop, but every lawyer will be a science lawyer.

THE FIELDS OF FORENSIC SCIENCE

The basic principle in examining physical evidence is rather simple. Whenever a person leaves a place, he or she takes something away from that place and leaves something at that place. It is the role of the forensic scientist and criminalist to examine what is left at the scene and hypothesize where it might have come from and who might have brought it there. Forensic scientific investigators are responsible for (1) identifying the bit of evidence, (2) comparing it with other similar substances and with substances known to be those of the victim or suspect, (3) individualizing the questioned substance as well as possible, and (4) using all these bits of evidence to help reconstruct the circumstances of the incident being investigated.

Specialized fields in the general area of forensic sciences include medicine (especially pathology), odontology (dentistry), anthropology, psychology and psychiatry, engineering, and toxicology. Ironically, many of these specialists do not work in the forensic or crime laboratory, but as medical examiners and as consultants to the medical examiner, the forensic lab, and the police.

The bulk of the work of a police forensic laboratory falls under the umbrella of *criminalistics*—the identification and best possible individualization of substances such as blood, hair, fibers, glass, paint, soil, plastic, markings, and prints. These labs also work on latent-fingerprint-removal techniques and fingerprint identification, firearms, tool marks, and questioned documents.

Each crime laboratory probably is organized a little bit differently, both internally and in its relations with the rest of the criminal-justice system. Some are attached to a medical examiner's office, others to a particular police jurisdiction, and others work on a consulting basis with regional law-enforcement agencies.

Many federal agencies have forensic laboratories, some of which specialize in certain areas. In addition to the FBI Lab, the lab of the Treasury Department's Bureau of Alcohol, Tobacco and Firearms gets high marks for research in firearms and tool marks. That lab also has a large database of inks and papers (not unreasonable because it is the investigative arm that looks into forgery of the country's money). In addition, the Drug Enforcement Administration (DEA) has several drug-testing laboratories.

Many states have regional labs that work with all local police departments. Others have one central lab. Some state laboratories are organized by county, and many large cities have their own crime labs. There are over 400 forensic laboratories in the United States, and more than 40,000 forensic scientists and criminalists.

POPULAR OPINION AND FORENSIC SCIENCE

In the past decade, the work of the medical examiner and forensic scientist has become a part of American popular culture. In almost every police drama, there is at least one episode a month where the police have a significant encounter with the scientific sleuths. These are often well-scripted segments that both move the story's drama forward and serve to teach the public about the work done in the back office, behind the work that officers and detectives do. Another way Americans have come to a deeper knowledge of forensics is through the reenactments on *The New Detectives,* a weekly show on cable television's Discovery Channel, which shows the work of forensic scientists, medical examiners, and criminalistic technicians.

Another development has brought the forensic scientist into American living rooms even more than police fiction or reenactments of how forensic science has solved old crimes: live coverage of major criminal trials. CNN and Court TV have done more to show people the work of forensic science in five years than have all the books ever written about the subject. No trial did this more than the 1995 trial of O. J. Simpson for the murder of his ex-wife and a friend of hers. For days, millions of Americans sat riveted to their sets, as prosecutors urged detectives and Los Angeles police lab forensic scientists to describe in minute detail the investigation they ran. Then the defense put on a slew of expert witnesses to cast doubt on the L.A. lab's methodologies and conclusions.

Fewer people followed the 1997 trials of Timothy McVeigh and Terry Nichols for the 1995 bombing of the federal building in Oklahoma City, but those who did were no less riveted by the forensic evidence presented there. The same is true of the 1997–1998 so-called Unabomber trial, or the trial of a number of Moslem radicals in the 1992 bombing of the World Trade Center in New York City. As American viewers have observed, many criminal prosecutions succeed or fail based on the quality of the forensic evidence provided to the courts.

PROBLEMS IN THE APPLICATION OF FORENSIC SCIENCE

In any field of scientific endeavor, the argument usually is made that science is value neutral—it is the use to which science is put that makes it good or bad. Forensic science, using the most advanced science and technology, has been shown time and again to be a great vindicator, as well as convicter. Despite this dual function, the vast majority of forensic scientists and criminalists work in police crime laboratories. As Joseph L. Peterson, a professor of criminal justice at the University of Illinois, writes in the American Chemical Society's book *Forensic Science,*

"Nagging questions remain about the objectivity of forensic science input to the judicial process and the quality of laboratory results and interpretations of evidence." Despite case law that police must provide defendants with *exculpatory evidence* (evidence that would tend to show innocence), Peterson and others argue that prosecutors often fail to show defendants all exculpatory forensic scientific evidence. They also argue that merely by deciding what tests will be performed, police and prosecutors can bias the results of forensic analysis.

While the tools of scientific crime solving and crime prevention—in fact, the tools of legal truth seeking—get better all the time, it is apparent that their use is not as good as it could be. In addition to the prosecutorial bias of most forensic evidence, both law enforcement and impoverished defendants often lack the training, personnel, and money to fully utilize the powers of forensic science, criminalistic techniques, and new technologies. In the mid-1970s, National Institute of Justice tests of crime laboratories found a majority to be deficient in their procedures; the institute cited slipshod laboratory procedures and failure to maintain a secure chain of custody for pieces of evidence. In fact, as of 1997, the FBI Lab was still not accredited by the American Society of Crime Lab Directors/Laboratory Accreditation Board (ASCLD/LAB).

Forensic science often is not much better in the private sector. In the case of DNA typing, before the FBI and state crime labs developed their own technology, private laboratories performed the majority of such procedures, and government officials spent much time determining how, if at all, to regulate it. As with each new scientific or technical procedure used by police and prosecutors, DNA typing has gone through a rigorous judicial review before being generally accepted as evidence in a criminal trial. Much of the argument about its admissibility has hinged on the question of how the courts can trust the findings of unregulated private laboratories.

Even after most DNA testing was taken over by state forensic labs and the FBI Lab, defendants have continued to question the quality of laboratory analysis. The questions about investigators' findings have also spilled over into the issue of drug testing. More and more employers are doing preemployment drug testing, testing for drug use when there is probable cause, and even random drug testing, despite the fact that independent researchers have shown a high and growing proportion of false-positive results and of careless or inappropriate laboratory techniques.

Even when forensic scientists follow flawless technical procedures, however, a far darker picture may still be painted of many criminal justice—and civil—uses of new technologies. As law enforcement assumes new abilities to investigate crime, it also assumes new abilities to infringe on the delicate rights that all U.S. citizens enjoy to be free from searches, seizures, and unwarranted government interference in their personal effects and communications. Further, these abilities to gather and combine thousands of bits of information about a person are available not only to law enforcement but also to governmental agencies that determine social welfare benefits, to potential employers, to those seeking credit references, and even to those merely selling soap.

The use of forensic science to match physical evidence at a crime scene to a particular suspect cannot be compared to its use in the creation of watch lists and behavioral profiles and the imposition of surveillance on those presumed to be possible suspects in some future crime. Nonetheless, such abuses of technology are increasingly possible, given the state of science and technology as we enter the twenty-first century.

As the science and technology of criminal investigation continue to move further beyond the crime laboratory—out of the hands of police and scientists and into the hands of bureaucrats, analysts, and politicians—the danger to our rights and liberties increases.

1

THE SCENE
OF A CRIME

In the middle of their second cup of morning coffee, on a warm summer Thursday in 2009, Detectives Johnson and Shipley are sent to the scene of a murder reported in a nearby middle-class neighborhood. As they arrive at a two-story single-family house, located on a quiet, shaded street, they are greeted by a uniformed officer who is standing at the base of the driveway. Another officer stands at the front door, and a third is on the walkway to the front door, speaking to neighbors.

The officer informs them that the victim is Mr. Anthony, white, 45 years old. His wife and children are at the seashore for the month; he was to join them this afternoon for a long weekend. The housekeeper arrived at 8:30 A.M.; a fourth uniformed officer is with her inside. Four more officers are on the way, and the officers on the scene have called for the medical examiner and the crime-laboratory technicians.

The housekeeper has told the uniformed officers that she let herself in the kitchen door around back with a key she took from inside the garage. Mr. Anthony's car was still in the garage; she assumed he was packing and getting ready to go to the beach.

Inside, she called for Mr. Anthony but got no answer. She looked through the house. The kitchen was tidy, the bed upstairs made. She then noticed that the television set was on in the den and thought Mr. Anthony probably was watching the morning news; she went to say hello and saw him lying on the floor, with blood all over, apparently dead. She called the police immediately.

While the uniformed officer is relating this to the detectives, the four other uniformed officers arrive—and an increasingly large crowd of neighbors gathers. The officers fan out, keeping the neighbors away from the front of the house. At the same time, they talk to the onlookers, seeking whatever information the observers might have.

Within a couple of minutes, the team from the medical examiner's office arrives—an assistant state medical examiner and two technicians. A step van, the state police mobile crime laboratory, also pulls up.

The detectives enter the house with the uniformed officer, the medical examiner, the two technicians with a stretcher, and the two state police crime-scene investigators. Before entering the house, all police and crime-scene investigators put on rubber gloves. This keeps them from leaving their fingerprints on anything they touch.

They also slip on clear plastic booties over their shoes. The officer inside speaking with the housekeeper has on gloves and booties, and he has put a pair of booties over her shoes as well. When each person leaves the house, he or she will take off the booties and put them in a plastic bag, marked clearly with the name of the officer, the date and the crime scene location. These will be examined at the crime lab for trace evidence. The booties are a new phenomenon throughout the country, an effort to keep investigative personnel both from contaminating the crime scene with material from outside and from leaving the crime scene with potentially valuable trace material.

The victim's house is a center-entrance colonial, with a stairway directly in front of the door, the dining room to the right, and the living room to the left. Behind the living room is the den, and a hallway runs under the stairway from the back of the living room to the kitchen. A half-bath is off the hallway near the den.

In the den, the medical examiner crouches next to the body. Detective Johnson stays with the medical examiner, while Detective Shipley begins to explore the house. The crime-scene investigators begin to examine the den and living room. Everyone works slowly and cautiously; the room is small, and they need to stay out of each other's way while still paying attention to every detail.

The dead man is dressed in casual cotton slacks and a polo shirt; he is barefoot. He is lying face down, head turned to the left. His body is positioned diagonally across the doorway. From the vantage point of the doorway, his feet are to the right, about 3 feet from the wall to the right of the door, while his head is to the left of the door, about 5 feet away from the wall. His head rests on an area rug that is on top of the den's hardwood floor.

The body is cold. There is a pool of clotted, drying blood around the head and neck. The medical examiner finds an apparent gunshot wound behind Mr. Anthony's left ear. This wound alone would have been enough to kill him, although evidence of other injury will be sought during an autopsy. The medical examiner's assistant puts a plastic bag around each of Mr. Anthony's hands; later, his fingerprints will be taken and compared with those in local police and FBI files, as well as the U.S. armed forces fingerprint registry.

There is no evidence that Mr. Anthony was tortured or sexually assaulted. There does appear to be a bruise on his left arm above the elbow—possibly, his killer grabbed him by the arm before or during the shooting. The medical examiner makes a note to examine Mr. Anthony's knees—he may have been forced to kneel during the killing. The angle of the bullet wound will help to determine whether this was so.

There is little the medical examiner can do at the scene. Back in her office and autopsy room, she will examine the body in minute detail, looking for evidence of injuries that were inflicted right before or even after death. She will remove from the body any hairs or fibers that do not appear to be the victim's. She will search for needle punctures. She will examine the victim's hands to see whether he either scratched or tore hair or fiber from his killer. Any skin under the fingernails or any hairs or fibers on his hands can be valuable evidence. She will also check to see whether the victim fired a gun recently.

The medical examiner will remove and weigh the victim's internal organs. She will open the skull and search for the bullet from the gunshot wound; if it can be found, it can be compared to bullets from any gun turned up by the investigation and suspected of being the murder weapon. She will draw blood from the victim, checking its type and matching it to any bloodstains on the victim's clothing or anywhere else in the crime scene. Any bloodstain that does not match the victim's blood could mean that the killer was injured. A number of other blood studies will be done; most important, to determine whether the victim had alcohol or other drugs in his blood at the time of his death.

The medical examiner backs away to give the crime-scene investigators more room and tells Detective Johnson that Mr. Anthony probably died around midnight, give or take a couple of hours. The position of the apparent bullet entrance and the fact that no gun was found imply that this death was not a suicide and not an accident, but murder.

Detective Shipley returns to the den and says that a preliminary look around the house shows no signs of forced entry and no evidence of robbery. There has been no ransacking of the master bedroom or, apparently, of the den. This evidence, added to the medical examiner's opinion, suggests that Mr. Anthony was killed by someone he knew, someone he had let into the house.

While the detectives and the medical examiner go out to the living room to continue their discussion, the crime-scene investigators meticulously examine the den. They photograph the room and the body dozens of times, from every angle, making sure to capture every item and surface in at least a couple of photographs. They also shoot video footage of the room, narrating as they go. This video is being sent live to technicians in the crime lab; a technician in the mobile crime unit outside acts as director, as if she were covering news or a sporting event. She asks one camera operator to scan the crime scene while asking the other camera operator to move in close to the bullet hole, the bloodstain and blood-spatter pattern, and finally the bruise near the elbow.

Meanwhile, various technicians are using a variety of fingerprint powders to dust every surface in the room, including the two glasses on the coffee table, a paperweight on the desk, the television, stereo equipment, and bookshelves. After the camera operators finish filming, they go back to the truck—removing and packaging their booties as they leave the house, then putting on a new pair as they reenter. They come back in with a large briefcase. In the case are a portable computer and a number of small devices. These are remote sensors, much like the ones used by scientists from the National Aeronautics and Space Administration (NASA).

Whereas the NASA scientists used such sensors to analyze in real time the surface of Mars, the crime-lab technicians are using them to transmit fingerprint images directly to the fingerprint examiners in the lab, who feed the information immediately into an automated fingerprint identification system (AFIS). AFIS is a computer-based expert system that automatically scans the points of identification on a fingerprint and produces a short list of the most likely possible matches from all the fingerprint databases the crime lab has access to.

By 2008, 37 states, the FBI, the armed forces, and Interpol have all provided links to each other for fingerprint databases. It is estimated that all states will be part of the network by 2012. It will be a matter of only minutes before potential matches are available to the fingerprint examiner. He or she will then print out the full sets of fingerprints that are possible matches, as well as the latent prints, and will do a visual examination of the latents and the possible matches.

Finally, the medical examiner's assistants remove Mr. Anthony's body, and the crime-scene investigators take samples of the bloodstains. The largest stain is on the area rug, but there are other spatter stains on the coffee table and the wall to the left of the door. All the stains are found in one direction away from the body, implying that a shot was fired from the opposite direction. There is no blood apparent in the part of the room where Mr. Anthony's feet were facing, and the crime-scene investigators have found no bloodstains in the living room or the rest of the first floor. From this, they assume that Mr. Anthony did not injure his killer significantly. They also have found no evidence that anyone tried to wash off blood in the kitchen or first-floor bathroom sinks.

Investigators find no discernible footprints on the hardwood floors of the living room, dining room, or den. Few implications can be drawn from this information, but it does suggest that the killer was not hiding in the garden, but probably came up the front walk. Mr. Anthony probably knew his killer and let him or her into the house.

Using magnifying glasses, the crime-scene investigators go over the den in careful detail, putting every hair, fiber, piece of paper, and other object they find into its own glassine bag for further examination. The hairs and fibers will be compared to Mr. Anthony's hair and the fibers from his clothing, as well as to hair from other people who frequent the house. Most fibers probably will be common, but there always is the chance that an odd fiber will turn up—from a particular rug or carpet or from a particular kind of clothing such as a fur coat—that family members can identify as belonging to someone they know. Scraps of paper may have writing on them or may have come from particular places that can be linked to particular people.

The two drinking glasses and a few other objects are removed from the coffee table for more careful fingerprint examination. There are no ashtrays and no smoked cigarettes, so there is no opportunity to try to match blood type to any saliva samples found.

The crime-scene investigators spend a few hours in the house, and so do the detectives. They interview the housekeeper in great detail, then go outside to ask the uniformed officers whether any neighbors think they have information about what happened the previous night: Did anyone hear anything? See anything? Notice an unusual car? Were there people hanging around in the crowd this morning that neighbors didn't know or recognize?

A neighbor and good friend has volunteered to call Mr. Anthony's brother, who lives in the next town, then go with him to the shore to get Mrs. Anthony and the children. She will call the detectives this afternoon so that they can arrange a time to speak with Mrs. Anthony. The detectives wait for the housekeeper's daughter to come and get her; she feels too upset to drive herself home. Finally, shortly after 1 P.M., they leave the Anthony house, stop at a deli to buy lunch, and go back to their office.

There, they begin to assemble a file on the case. They write their own reports and get copies of the reports submitted by all the uniformed officers—more detailed versions of what they had been told in their informal interviews of neighbors on the Anthonys' front lawn. Names, addresses, and telephone numbers of neighbors and possible witnesses are recorded so they can be reinterviewed in greater detail.

Within a couple of days, the medical examiner will send her report, stating her professional opinion about the cause and time of death. The photographs and room diagrams will arrive, as will the preliminary laboratory reports on the blood type of the stains, whose hairs and fingerprints were found, and where the pieces of fiber and paper might have come from. More detailed toxicology studies of the victim probably won't be available for a week or so.

Matches of Mr. Anthony's fingerprints will reveal whether he had a criminal record, either locally or anywhere else a police jurisdiction contributes criminal and fingerprint files to the FBI's national computer database network. The detectives will learn whether Mr. Anthony had served in the military, whether there were any civil-court proceedings against him, his credit rating, and a host of other information available to those who know an individual's Social Security number.

In short, using a combination of forensic evidence, interviews with those who knew him, and database information, Detectives Johnson and Shipley will find out who Mr. Anthony was, what he did, and who might have wanted to kill him.

While they eat their sandwiches and drink their sodas, they take a few minutes to relieve the tension they always feel—despite years of experience—upon going to another murder scene.

After lunch, and after they finish their reports, they will return telephone calls they received while they were out. Most of the calls either are providing information on one of the six murders they are currently investigating or are asking for information about one of the trials being prepared by state prosecutors using information they gathered in one of the more than 40 cases they investigated that are awaiting trial. They see no end in sight: Tomorrow, or the next day, they will get another call, about another murder.

2

VICTIMS WITHOUT IDENTITIES, MURDERS WITHOUT BODIES

At 7:30 P.M. on July 15, 1988, Donald Santos dropped his 26-year-old wife, Mary Rose, at the commuter-bus terminal in downtown New Bedford, Massachusetts. Mary Rose was going to walk a few blocks to see a friend and planned to call her husband for a ride home later. He never heard from her again.

Mary Rose Santos was last seen dancing at the Quarterback Lounge at 1 A.M. on July 16. "Right now, the five-year-old asks, 'What time is mommy coming home?'" Donald Santos told the New Bedford *Standard-Times* a few days later. The couple's older son, age seven, thought his mother had left because he had been bad.

The disappearance of Mary Rose Santos, a member of New Bedford's large, mostly poor and working-class Portuguese community, did not raise the suspicions of many people, except her family. The New Bedford police dutifully took the missing-person report, one of many they take every year. There are not enough police officers anywhere in the country to follow up on missing-person reports thoroughly, and this case was no different.

By the time what was left of Mary Rose Santos's body was found, on March 31, 1989, New Bedford police were taking reports of missing persons more seriously, especially if they were young Portuguese women who had drug problems or worked as prostitutes in the city's Weld Square area. Mary Rose Santos's remains were the eighth of nine sets found by the spring of 1989, which New Bedford and Massachusetts state police believed were all victims of the same killer or group of killers.

The first body in what was to be called the Highway Killings was found on July 3, 1988, just 12 days before Mary Rose Santos danced into eternity. The partially mummified remains of a young, petite woman were found about 100 yards off Route 140 in Freetown, a few miles from downtown New Bedford. A few weeks later, on July 30, a motorist stopped by the side of Interstate 195 near the Reed Road exit to relieve himself, and he discovered the skeletal remains of another young woman.

The *Standard-Times* reported these two finds in short, un-bylined articles, and the newspaper's librarian, Maurice Louzon, opened a file headed "skeletal remains." On August 13, the paper did its first extensive follow-up on the two bodies, focusing on the investigation by Massachusetts state trooper William Delaney, who was assigned to the office of the Bristol County district attorney, Ronald Pina, to investigate homicides.

Delaney told the reporter that because of the way the first body's bones and skin were preserved, investigators theorized that the victim had died just before the onset of cold weather the previous fall, about nine months before the body was found. The woman was petite, probably in her late teens or early 20s, had had her jaw wired from a previous break, and had had other dental work. She had been found lying on her back, and there were no signs of a struggle at the scene.

The second body was that of a woman who was older than the first—between 30 and 40. It was an intact skeleton, found lying on its back. She probably had been killed in the spring; again, there were no signs of a struggle at the scene of discovery.

Delaney had published the descriptions of both women nationally and had dismissed about 20 responses from as far away as Florida. He had obtained a computer listing of 1,724 Massachusetts women who were missing and who were between 4 feet 11 inches and 5 feet 5 inches tall, the approximate height of the two victims. Next, he planned to cull that list for possible matches and to request dental records of those possible matches. He told the reporter that the chances of matching the skeletons to any known missing person were slim at best.

There are an estimated 5,000 sets of skeletal remains found each year in the United States, and they are among the most difficult cases for police and medical examiners to solve. With a relatively fresh homicide, clues usually are available in abundance, although making them tell their grisly tale with enough clarity to ensure a conviction is not easy. If there are no eyewitnesses and no murder weapon is found, and police do not find a suspect within the first few days after a killing, solving murders is very difficult. With skeletons, very little about the time and manner of death is obvious, and the time lag between the killing and the discovery makes solving the case that much harder. Nonetheless, forensic anthropologists have been able to use skeletal remains to determine that murder is also an ancient phenomenon. Investigators have found skeletons of early humans for whom it can be determined that their caved-in skulls were caused by repeated blows from an object and not just by random head trauma. Also, in the peat bogs of Northern Europe, the so-called bog mummies show that 10,000 years ago, some individuals became the victims of ritual murders, and their bodies were simply left in the peat bogs, rather than being buried according to the custom of the day.

Although skulls and skeletons can reveal gross evidence of some violent deaths, they rarely allow for precise determinations of the time and cause of death. This makes the work of medical examiners—whose job is to detail the time and cause of death as exact as possible—much more difficult. Only if there is obvious trauma to the bones caused shortly before or during the killing is there enough information to pinpoint a cause of death. Violent deaths, even the most gruesome, usually involve injury only to soft tissues and organs, which decompose or, if a body is left in the open, are destroyed or removed by scavenging animals. With the first two bodies found in the New Bedford killings, the county medical examiner did a microscopic examination of the bones to see whether he could find any bullet or knife nicks or any new minute fractures, which sometimes are enough to determine a cause of death. He could not.

Without the soft tissue with which to estimate time of death by the degree of decomposition, telling how long remains have been at a site often is relegated to guesswork. Even with remains that include some soft tissue, it is difficult to pinpoint the time of death. In the case of the first New Bedford body, investigators and medical examiners theorized that the partial mummification had occurred because the body was left in the open during the autumn, when the days are sunny but cool and the nights often are frigid. At that time of year, the decomposition that occurs each day becomes solidified during the night and, although the body's water content is quickly siphoned off, the skin and muscle are not decomposed rapidly, as happens during hot weather—and the summer of 1988 in southeastern Massachusetts was indeed hot. Therefore, they expected a time of death after the hot summer, during the autumn of 1988. The investigators were surprised by later findings. When the remains were identified on December 8, using dental records, they turned out to be those of Debra Medeiros—who had disappeared on May 27, 1989, only six weeks before her remains were found.

More than just the state of the remains themselves can help investigators to pinpoint how long a body has been in the open. Clothing decomposes at different rates, the state of the soil under the remains can tell how long they have been there, and even the insect life in the soil can help to determine the length of a body's stay in the open.

While skeletons often yield only the most minute clues about the time and manner of death, persistence and some luck often can lead to positive identification of the remains. If the records of those who are missing are good enough, forensic anthropologists and odontologists (forensic dentists) can identify remains with remarkable accuracy. Dental records provide the best identification, followed by X rays of injuries that have healed.

Working with as little as a handful of bones and the crowns of one or two teeth, forensic scientists have identified the victims of disasters new and old, homicide victims, those who "disappeared" during the brutal reign of terror in Argentina, and victims of "ethnic cleansing" in the former Yugoslavia, Rwanda, and other countries. In a celebrated case, forensic scientists were instrumental in identifying the notorious Nazi Angel of Death, Josef Mengele. In addition, by using both forensic odontological and anthropological techniques and DNA samples, investigators have been able to match the tiniest remains to almost all of the victims of major disasters in the 1990s, including the ValuJet crash in the Everglades, the TWA explosion off the coast of Long Island, and the Oklahoma City bombing.

When skeletal evidence is all that remains, each of the body's 206 bones and 32 teeth can tell a story about the life of its former owner—that person's *osteobiology*. Injuries are recounted through healed fractures. Illness can be inferred from such deformities as bowed legs (a sign of rickets), the telltale shortness of a polio victim's limbs, the skeletal damage caused by such diseases as syphilis or tuberculosis, or by bone infections such as osteomyelitis. Childbirth is also detected from the way the bones that form the pelvis are set in some women. The injuries involved in skull fractures can also reveal potential controversies in forensic interpretation. For instance, in the 1997 trial of Louise Woodward, the British nanny convicted of killing an infant in her care in a Boston suburb, experts in child abuse and forensic anthropology argued about whether the death was caused by a single act of violent shaking or as an artifact of months-old abuse.

Skeletal evidence can sometimes even reveal occupational information. That is, a person's occupation can change the skeletal structure. For instance, waitresses and tennis players show signs of their arm strength in their bones; their strong side is more developed than their average side.

The male pelvis differs from the female pelvis—narrow and steep for a man, broader and shallower for a woman, regardless of childbirth. The heads of long limb bones also differ—men's are wider—and there are a host of other sex-related features.

Racial features also differ. Forensic anthropologists often use features of the eye sockets and nose to categorize people in one of three racial groups: Mongoloid (broadly defined as Asian), Negroid (African), and Caucasoid (European). In Negroids and Mongoloids, the ridge of the nose often is broad in relation to height; in Caucasoids, it is narrower.

The skull also gives clues to a skeleton's age. The *basilar joint* (where the two main bones of the skull's underside meet) fuses in the late teens. Another clue to age is in the long bones of the arms and legs. For instance, where the knob and shaft of the femur (the large upper leg bone) meet any areas of cartilage, the degree of calcification indicates maturation. Incomplete calcification signals that full maturity has not been reached. Once the skeleton is fully adult, microscopic analysis of bone degeneration is necessary to try to pinpoint age.

The body's 12 long bones—two each of the femur (thighbone), tibia (shinbone), fibia (calf bone), humerus (upper arm bone), radius (bone on the thumb side of the forearm), and ulna (bone on the little-finger side of the forearm)—are all durable and often are found even in minimal skeletal remains. By measuring long bones, physical anthropologists can approximate the living individual's height by using the Trotter formula, developed by Mildred Trotter, a professor of anatomy at Washington University in St. Louis. Trotter spent time after World War II working for the Army in Honolulu, helping to prepare U.S. servicemen who had been buried overseas for final burial in the United States. In exchange for her service, the government let her do research. Using the long bones of hundreds of servicemen, Trotter figured out algebraic formulas for the relationship between bone length and the height of a living subject. She found that the correlations were different for different races and required different formulas and also that the bones of the legs were far more reliable in determining height than were the bones of the arms.

For the remains of a white male, physical anthropologists can use the following formulas to estimate height from one long bone:

2.38 (femur length) + 61.41 centimeters
= height in centimeters ± 3.27 centimeters

2.68 (fibula) + 71.78 = height/cm ± 3.29 cm

2.52 (tibia) + 78.62 = height/cm ± 3.37 cm

3.08 (humerus) + 70.45 = height/cm ± 4.05 cm

3.70 (ulna) + 74.05 = height/cm ± 4.32 cm

3.78 (radius) + 79.01 = height/cm ± 4.32 cm

If the femur and tibia are both present, the margin of error can be reduced even further:

1.30 (femur + tibia) + 63.29 = height/cm ± 2.99

In the case of remains of a teenage boy found with only a 46-cm femur present, the Trotter formula would estimate the young person's height at

2.38 (46) = 109.48 + 61.41 = 170.89 ± 3.27
(i.e., between 167.62 cm and 174.16 cm; at 2.54 cm

per inch, between 65.99 inches and 68.57 inches—
i.e., between 5 feet 6 inches and 5 feet 8½ inches).

Although this information is nice to know when trying to put a name onto a single set of bones, it is vital when trying to match remains with people assumed to be at the scene of a disaster. In her book *A Plane Is Missing,* author Susan Sheehan describes the meticulous care given by Tadeo Furue, the Japanese-born anthropologist at the U.S. Army's Central Identification Laboratory in Honolulu. In 1982, Furue identified all 22 men who had been aboard a B-24 that crashed into the side of Mount Thumb in Papua New Guinea on March 22, 1944. Furue made extensive use of the Trotter formulas to determine the approximate height of each partial skeleton he assembled from the commingled remains taken from the wreckage nearly four decades after the plane crashed.

Once Furue had sorted the bones onto 22 stretchers—a task that took weeks—he used the Trotter formulas to assign height. Then he compared the sets of bones to the airplane's manifest. Because many of the sets of remains were approximately the same height, and because most of the men on the manifest were of about the same age—almost two-thirds were between 20 and 25—Furue needed to go many steps farther to positively identify the men.

Furue turned to another anthropometric formula, called the Rohrer body-build index, to further his effort. The index, devised in 1921 by the German physical anthropologist Fritz Rohrer, takes the weight in grams divided by the height in centimeters cubed and multiplies the quotient by 1,000 to come up with a body-build index that is between 1.00 and 1.67. For instance, I am 6 feet 2 inches tall and weigh 200 pounds. My body-build index is

200 pounds = 90.9 kilograms (90,900 grams)
74 inches = 187.96 centimeters

187.96 × 187.96 × 187.96 = 6,640,443,150

90,900 divided by 6,640,443,150 = 0.00153

0.0012298 × 1,000 = 1.53

I fall within the "heavy build" portion of the Rohrer body build index (1.00 to 1.15 is slender, 1.15 to 1.40 is medium, 1.40 and up is heavy).

By comparing the military records of each crash victim to the sets of remains, Furue was able to narrow the possibilities for each. Finally, he turned to dental records and compared them to the teeth found—in some cases, as few as two or three; in other cases, nearly whole jaws. Using all the evidence from his many tests, Furue was able to positively identify all the skeletons.

In the case of an airplane that crashed almost 40 years earlier, Furue's meticulous exercise may be merely academic. In contrast, when an airplane crashes with hundreds of dead and no survivors—and the passenger list may not be 100 percent accurate—the ability to make positive identification is imperative so that families can make life-insurance claims and claims against the airlines and manufacturers of the equipment. Positive identification also may help investigators to determine who may have checked in luggage that contained a bomb and then walked away from the plane.

Stanley Schwartz, a professor of dentistry at Tufts University in Boston, began his work in forensic odontology in just this way when he volunteered his services after the crash of a Delta airliner at Boston's Logan Airport in July 1973. Schwartz, one of about 80 forensic odontologists in the country, later became Massachusetts's official dental examiner.

In the fall of 1988, with more than 15 years of forensic experience, Schwartz began poring through dental records of missing women to try to identify the New Bedford bodies. He began working in mid-November, between the time the third body was found, on November 8, and the fourth was found, on November

29. The November 29 find and a fifth, on December 1, were made after an extensive search was carried out by Massachusetts and Connecticut state police, using dogs specially trained to find dead people and animals.

In just a few days, Schwartz's work paid off. On December 5, he identified the fourth body found as that of Dawn Mendez, 25, the mother of a five-year-old son, who had disappeared on her way to a baby christening September 4. A partial fingerprint also helped to identify Mendez, who had a police record. The next day, Schwartz identified the second body as that of Nancy Paiva, a 36-year-old heroin addict and mother of two who had disappeared July 7. Two days later, he identified the first body found as that of Debra Medeiros, 29, of nearby Fall River, Massachusetts, who had been missing since she left her boyfriend's new Bedford home on May 27.

At first, investigators had thought that the third body was Paiva because it was wearing Paiva's clothes, but Schwartz's dental identification proved investigators wrong.

Schwartz stayed busy on the New Bedford cases. Over the next three months, he identified the three other bodies that had been found. The body found November 8 was that of Debra Greenlaw DeMello (it took Schwartz a solid week of poring over dental records to identify her); the body found December 10 was that of Rochelle Clifford Dopierala, a Cape Cod resident. On March 1, 1989, Schwartz identified the last of the three bodies, found on December 1, as Debroh McConnell, of Newport, Rhode Island.

Unfortunately, McConnell's was not the last body of the Highway Killings collection. When the ground began to thaw, state police, who now believed there was a pattern to the killings, once again called specially trained dogs and their handlers from all over New England to conduct an extensive search of the roadsides of Interstate 195 and Route 140 near New Bedford. There were about a half dozen young women from the area, mostly drug addicts, as well as some prostitutes, who remained missing. They found another body on March 28, 1989; the next day, Schwartz matched the body's teeth to the dental records of Robin Rhodes, a troubled 29-year-old mother who had left her 7-year-old son with her parents the previous April and had disappeared. Her family didn't report her missing until after she failed to come back for both her own birthday and her son's, in June.

Three days later, two children found Mary Rose Santos's body, and again Schwartz identified it within days. Another three-day search by the dogs in the middle of April was fruitless, but on April 24, the ninth body was found—that of Sandra Botelho, a mother of two who had been missing since August 11, 1988.

By the time the bodies were found in the spring, police had a pretty good idea about to whom the bodies would turn out to belong. There had been a spate of missing women in the summer and fall of 1988; after all nine bodies had been found, District Attorney Pina said that he would not be at all surprised if two more bodies turned up—belonging to Christine Montiero, a heroin addict who had disappeared May 16, 1988, and Marilyn Roberts, the 34-year-old daughter of a New Bedford police officer, who was reported missing in December 1988, after having not been seen for months.

When the police interviewed numerous prostitutes and other women in the Weld Square area of the city, a number of prostitutes told the police that they had been attacked in the summer and fall of 1988 by a man who said he was in law enforcement. He picked up the women, and once they were in a quiet location, he tried to strangle them. One prostitute said that a man in a white pickup truck had raped her one highway exit from where three bodies were found. On December 12, Neil Anderson, 35, was arrested and charged with the rape; he immediately became the prime suspect in the killings.

In early 1989, District Attorney Pina impaneled a grand jury to hear evidence and told the news media that he had narrowed the list of suspects to four men. One was Anderson. A second was a former New Bedford lawyer, Kenneth Ponte, 39, also a county deputy sheriff who served civil-process papers. Ponte had been seen with or was known to have connections with four of the six dead women: Medeiros, Dopierala, Mendez, and Rhodes. Robin Rhodes's mother recalled that her daughter had said a rich lawyer was going to help her kick her drug habit; Ponte had moved to Florida in October about the time that women stopped disappearing from the streets of New Bedford. A friend of Ponte's said that he often "took in strays."

Initially, Ponte said that he would allow investigators to take samples of his saliva and of his head and pubic hair, but he later refused, and a court declined to order that the samples be taken. Ponte also refused to allow investigators to take photos of a tattoo on his arm. He had been convicted several times on drug charges in the late 1960s and early 1970s and had been jailed in 1971. In 1975, he was given a pardon by then Governor Francis Sargent so that he could become an attorney and a member of the Massachusetts bar.

Pina continued to use the grand jury as an investigative tool, bringing dozens of witnesses before it in March, April, and early May. In early March, he told the media that he had narrowed the suspects to two. In mid-March, he received a two-page, typed letter from "a citizen" who said that he or she had "quite a bit of information about missing women" and that a body should be found in the vicinity of Horseneck Beach, a few miles from where any other body had been found. On Friday, March 31, two children did indeed find Mary Rose Santos's remains where the anonymous letter-writer said they would be.

On Tuesday, May 2, Margaret Medeiros, a 22-year-old prostitute and drug addict, told that grand jury that she had been picked up in the summer of 1988 by a man driving a Ford Bronco or similar truck and that he had tried to break her neck, telling her, "I'm going to do to you what I did to the other bitches." She said that at least 14 other women she knew who worked the Weld Square area had had similar run-ins. She said that the man tried to pick her up again in late April 1989 and allegedly attacked her two roommates that same April night.

Medeiros told reporters that she identified the man for the grand jury from a photo lineup. She had been shown photographs of three men: her attacker, a Rhode Island man, and Neil Anderson, the rape suspect arrested in December.

The *Standard-Times,* which refused to identify the man because he had not been charged, did say that its library files included articles about him. He was a resident of East Freetown and had been arrested three times previously for rape but was never convicted.

There have never been any arrests in the New Bedford Highway Killings case.

Nonetheless, Schwartz was lucky in his victim-identification efforts: He had relatively complete sets of teeth to examine, a set of probable victims, and fairly good dental records for most of those women. Not every forensic examiner is so lucky, but a Connecticut case shows what can be done with even the most meager evidence.

On the night of November 18, 1986, a 39-year-old former flight attendant, Helle Crafts, disappeared. Six weeks later, Connecticut state police and crime-scene experts from the Connecticut State Police Forensics Laboratory, acting on information from witnesses, began searching a riverbank miles from the Crafts's home in Newtown. Over the next 18 days, they made a minute examination of the crime scene off River Road in Southbury, on the banks of the Housatonic River. Their investigation of the riverbank, as well as the Crafts's house, led them to hypothesize that Mrs. Crafts had been the victim of a brutal murder and that her body had been disposed of in a most gruesome way.

Helle Crafts, police speculated, had been murdered by her husband, Richard, then 48, in their Newtown home on the night of November 18 or early in the morning of November 19. On November 21, police said, after cutting his wife's body into a number of parts with a chain saw and putting the parts in plastic bags, Richard Crafts drove to the banks of the Housatonic in a rented truck and, using a rented wood chipper, proceeded to mulch his wife's body.

Witnesses remember seeing a man using a wood chipper on the riverbank that day, despite a driving snowstorm. Also, the Crafts's babysitter remembered seeing a huge stain on the carpet in Mr. and Mrs. Crafts's bedroom shortly after Helle Crafts disappeared. Richard Crafts told investigators that he got rid of the carpet soon after his wife disappeared because he had spilled kerosene on it.

During nearly three weeks of probing, investigators found various minute human remains among the wood chips on the riverbank, including pieces of bone, a fingernail and a toenail, teeth and dental crowns, a finger, 21 clumps of hair containing over 2,300 hairs and blood, and pieces of plastic from several different plastic garbage bags. They also found a chain saw, owned by Richard Crafts, in the river.

Over the next few months, Henry Lee, chief of the Connecticut State Police Forensics Laboratory, conducted over 52,000 experiments on 251 items recovered by the state police. Lee also tested bloodstains on a mattress taken from the Crafts's bedroom.

Richard Crafts was charged with his wife's murder and tried beginning in April 1988. Twenty-eight days into the trial—which eventually ended in a mistrial when one holdout juror refused to deliberate any longer—prosecutors launched a parade of expert scientific witnesses whose testimony described in minute detail the process of trying to identify a person from the slightest possible remains.

On Monday, May 10, 1988, the state of Connecticut began to

lay out its forensic case. R. Bruce Hoadley, a University of Massachusetts professor of forestry and wood technology, testified that marks on wood chips found on the bank of the Housatonic matched the marks on wood chips found in the wooded lot on the Crafts property and in the back of a truck that Richard Crafts had rented about the time his wife disappeared.

The next day, Albert Harper, a biological anthropologist from the University of Connecticut, told the jury that 69 bone fragments he examined from the riverbank mulch pile were pieces of bone from a mammal, many of them definitely from a human adult. Because of their small size, he was unable to determine age, sex, or race. "I've never seen anything like this before," Harper said of the specimens he examined. "The only thing I've ever seen close to it was a cremation. Oftentimes skeletons are broken on the ground, but these were very small pieces," Harper said, noting that the edges of many of the bones were straight, but not smooth, and that they had been cut by something straight, but not necessarily very sharp.

Next, C. P. "Gus" Karazulas, a forensic odontologist, testified that by taking hundreds of X rays at all angles, he had been able to make a positive identification of one tooth and one gold/porcelain crown as belonging to Helle Crafts. Karazulas testified that the tooth was ripped from Crafts's mouth with "traumatic force," tearing jawbone with it. Lowell Levine, a forensic odontologist for the state of New York, concurred, saying of the tooth in evidence, "The lower left second bicuspid belonged to Mrs. Crafts when she was alive."

How could one tooth and some dental work lead to a positive identification? Each tooth has five visible surfaces, as well as the root, and trained odontologists can distinguish nearly 100 tooth characteristics, both on the tooth itself and on X rays. A comparison of post-mortem X rays to those in a dentist's file is the technique most often used for identification. In addition, forensic dentists are able to match the characteristics of dental work to the often meticulous records now kept by many dentists.

Even damaged teeth can tell the trained eye a significant story. The arrangement of teeth, the placement of fillings, missing teeth, the alignment of teeth in the jaw, even the condition of the gums, can help the forensic dentists. The metal content of fillings, crowns, and other dental work can be determined in the laboratory and can help to tell how old the dental work is. Also, the pulp inside a tooth can help to determine a victim's age.

In 1986, the American Dental Association began its dental data-disk program, in which dentists bond a nearly indestructible disk one and one-half times the size of a pinhead onto an upper molar. Under an ordinary magnifying glass, the 12-digit code can be read and there is a 24-hour hot line for its use. Those who developed the technology hoped that, over time, all U.S. citizens will be dentally encoded early in life for later identification, if necessary. As late as 1998, however, few dentists were routinely encoding their patients.

In addition to identifying remains, forensic dentists can identify both victims and perpetrators of crime through examination of bite marks. Perhaps the most famous bite-mark evidence came in the 1980 trial of Theodore Robert "Ted" Bundy in Florida. On January 8, 1978, Bundy arrived in Tallahassee, Florida, after escaping from a Colorado jail where he was being held for murder. He had recently been convicted of kidnapping in Utah, and police suspected him in the murders of young, pretty women with long hair in four states: Washington, Oregon, Utah, and Colorado. It seemed that wherever Ted Bundy was, beginning in about 1973, young women began dying on and around college campuses in the vicinity.

In the early morning hours of January 15, it happened again, at the Chi Omega house at Florida State University. At 3:00 A.M., one of the sorority sisters saw a man run down the stairs through the foyer and out the front door, a man wearing a stocking cap and carrying a piece of wood or a club.

Upstairs, two women were dead and two were severely injured, the victims of savage attacks that had taken place over a period of 25 minutes while they slept. Survivors Kathy Kleiner and Karen Chandler both had skull fractures, broken jaws, broken teeth, and severe cuts on their heads, faces, and torsos. Blood and pieces of wood bark were splattered about the room.

Margaret Bowman and Lisa Levy were not so lucky. Bowman had been clubbed to death; pieces of bark were stuck to her hair and face by blood. Lisa Levy apparently had been strangled. Her right nipple had been bitten nearly off, and there was a deep bite mark on her buttock. This bite would, in effect, convict Ted Bundy and, nearly nine years later, send him to Florida's electric chair.

Richard Souvion, Florida's chief forensic odontologist, used enlarged photographs of the bite mark on Levy's buttock and of Bundy's teeth, superimposed on each other on a screen, to show the jury that, within reasonable dental certainty, those teeth had caused that mark.

While even the naked eye of a juror can see similarities in bite marks and teeth, the forensic dentist uses an array of equipment, including infrared and ultraviolet photography, electron microscopy, and computer analysis to make precise comparisons of teeth. Such evidence can prove quite compelling. For instance, a forensic odontologist was the person who finally bit a hole in Marv Albert's defense against assault charges. More important even than the testimony by a hotel VIP-relations manager who witnessed Albert's prancing about his hotel room in women's underwear was the testimony of a forensic odontologist. This expert linked Marv Albert to bite marks on the back of a woman who alleged that he had forced her to engage in oral sex on him, thereby leading Albert to stop his 1997 trial and plead guilty to assault, causing him to lose his job as one of the most prominent sports announcers in America.

For victims of homicide, the changes in tissue around a bite on a corpse also can help to pinpoint the time of death. Biting the victim is common in violent crimes. In 1977, Los Angeles police found identical bite marks on a number of victims of the serial killer who would become known as the Hillside Strangler.

Often, bite marks are also important in sorting out child-abuse cases where parents, friends and acquaintances, and babysitters are all logical suspects.

Obtaining dental impressions of suspects requires only a search warrant today, despite the cries of civil libertarians that it is a violation of both the Fourth Amendment right against searches and seizures and the Fifth Amendment right against self-incrimination (the same is true of samples of bodily fluids such as blood, semen, and saliva). Courts in more than half the states and on the federal level have ruled that where there is probable cause, investigators have a right to try to match scientifically valid evidence by taking samples from suspects.

Victims who know a little about the possible forensic value of bite marks—they can stay vivid enough to analyze for weeks—may bite an attacker to help in later identification, as well as struggling in other ways.

In addition to routine criminal and disaster investigations, forensic dentists and anthropologists are increasingly involved in politically charged cases. The American Association for the Advancement of Science (AAAS), through its Science and Human Rights Program, along with human-rights groups such as Americas Watch and Helsinki Watch, is turning to forensic scientists to do objective analysis of alleged human-rights violations. In 1976, the program began to monitor abuses of scientists and dissidents, and a number of pathologists and anthropologists have traveled throughout the world on AAAS missions, both to identify remains of possible human-rights victims and to train scientists in forensic pathology and anthropology.

Throughout the late 1980s and the 1990s, as country after

country has moved from dictatorship to democracy, teams of forensic scientists, pathologists, and anthropologists, mostly from the United States, have worked to help commissions, police departments, and courts identify the victims of political murders, many of whom were buried in mass graves, in such disparate places as Chile, Guatemala, Brazil, Bosnia-Herzegovina, Rwanda, and the Philippines.

In May 1988, Jorgen L. Thomsen, a pathologist at the University Institute of Forensic Medicine in Copenhagen and president of the Committee of concerned Forensic Scientists and Physicians for the Documentation of Human Rights Abuses, visited El Salvador for AAAS to help authorities determine whether two young men had been tortured and murdered, possibly by the military.

The AAAS has also worked with the Argentine National Commission on the Disappeared to try to determine what happened to the more than 9,000 people who "disappeared" during the brutal government crackdown on alleged dissidents, which began in 1976. Clyde Snow, a former Federal Aviation Administration (FAA) anthropologist who became the country's foremost freelance forensic anthropologist in the mid-1980s, led a team that trained Argentinean students in the rudiments of the art; in 1989, those students still were sifting through skeletons in mass graves throughout that country.

Snow, who is based in Norman, Oklahoma, also was involved in the identification of Josef Mengele, the Nazi Angel of Death, who experimented on concentration-camp inmates during World War II. Mengele escaped post-Nazi Germany and apparently lived in South America for more than 35 years. As Nazi hunters, including the famed Simon Wiesenthal, were closing in on the elusive Mengele in early 1985, rumors began to surface that he had drowned and was buried at Embu, Brazil. A German couple, Wolfram and Lisolete Bossert, led São Paulo police to the grave— where, they said, Mengele had been buried under an assumed name in 1979.

An international team of forensic experts examined the remains in the Embu grave—a fairly intact skeleton including long bones and a skull, six teeth, a few clumps of hair, and even some rotting clothing. Slowly, over the course of weeks, the clues began to show that the bones at Embu were Mengele's. First, the forensic scientists determined that the skeleton was that of a Caucasian male and that his height was within the bounds of Mengele's—5 feet 8½ inches, by the account in his SS file. Mengele's dental records, however, were incomplete and not very precise, making an examination of the teeth less useful. There was no X ray of the hip fracture that SS records referred to, although such a fracture was evident in posthumous X rays of the Embu remains.

A microscopic count of the blood-carrying canals in the skeleton's femur—the more canals and the more fragmented they are, the older the individual—was conducted by Ellis Kerley, an anthropologist in Maryland, who showed the skeleton to be between 64 and 74 years old; Mengele would have been 69 the week after his alleged swimming accident. The nerve canal along the top of the mouth toward the front teeth showed up on X rays as being wide, matching reminiscences and photos of Mengele's gap-tooted smile. As Snow told a reporter for *Science 86* magazine months later, the team's identification was somewhere between "probable" and "highly probable."

Then came Richard Helmer, a German pathologist. In the mid-1970s, he developed a process of photographic superimposition by which he can compare skulls to photographs of possible matches. Helmer calculates a number of reference points on the skull, such as the placement of the nose opening, eye sockets, cheekbones, and eyebrow ridges. Then, using anthropometric tables of the thickness of muscle, fat, and skin for people of various ages and races, he marks about 30 points on the skull with pins and dabs of clay. These tables were developed by nineteenth-century anatomists, who took the measurements by putting cork on the ends of pins, pushing the pins into the faces of cadavers, and recording how far the cork was pushed along the pin when the pin finally hit bone.

Using two high-resolution cameras and an image processor, Helmer records pictures of the skull with its reference points, as well as photographs of the possible match—as many as possible—then moves the images along a track until the photo of the living subject is superimposed over the skull.

For the Mengele identification, Helmer used both a formal portrait from 1938 and a number of photos that the Bosserts took of the man they believed was Mengele shortly before his death. In each case, there was a match sufficient that by the end of Helmer's two-day investigation, the team was willing to say, within scientific certainty, that the bones at Embu were indeed those of Josef Mengele. Helmer has since used his superimposition technique in Argentina under the auspices of the AAAS.

Betty Pat Gatliff, a medical artist who often works with Clyde Snow, uses a technique similar to Helmer's to do a three-dimensional sculpted reconstruction of a face from a skull. Photographs of the facial reconstruction then are used in an effort to identify the skull remains. Similarly, Karen Taylor, the forensic artist for the Texas Public Safety Commission in Austin, uses the same standard tables and landmarks to create sketches from skulls. People who believe the sketch matches a missing loved one send in photographs taken as close to the time of disappearance as possible, and the photo is superimposed over a photo of the skull that has been found.

Readers of crime fiction will remember the novel *Gorky Park,* in which the partially decomposed skulls of three people killed in Moscow's Gorky Park were taken to an eminent scientist, who turned insects loose on the skulls to finish eating the flesh away before he began his reconstruction.

In the United States, computer-graphic techniques have carried the identification process many steps farther than Helmer's and other manual techniques. Two medical illustrators from the University of Illinois at Chicago—Scott Barros and Lewis Sadler—have developed computer software that does in minutes what had previously taken hours to do by hand: create an age-adjusted picture of a child. Barrows and Sadler have been creating these so-called "age-progression" drawings since the early 1980s, as an outgrowth of Sadler's work in projecting the outcome of cosmetic surgery, and in response to a television producer's request for a portrait of what a missing child might look like years after being kidnapped.

Barrows and Sadler used their knowledge of anatomy—the face is defined by the shape and size of 14 major bones and more than 100 muscles—to develop a way to project what a child's face might look like as it matured. Their portraits have proved stunningly accurate. They first used the technique on the pictures of Kathleen and Deborah Caruso, sisters who had been abducted seven years earlier by their father in a custody dispute. Within 20 minutes of broadcasting the age-progressed pictures on national television in April 1985, the sisters' neighbors and school teachers in Kettering, Ohio, were able to notify authorities, who then arrested the father and reunited the girls with their mother.

Since 1985, Barrows and Sadler have created dozens of other age-progressed portraits. They developed a computer program in 1987 as a way to speed the process and cut into the backlog of more than 7,000 cases on file at the National Center for Missing and Exploited Children.

To project what a youth's face might look like based on a photograph of the child taken years earlier, Barrows and Sadler first look at 48 facial landmarks, such as the corners of the eyes and nose. Then they measure distances between these landmarks. After determining from medical research the rate of growth of facial features over the period of time in question, they recalculate these distances and create the age-progressed portrait.

By hand, this is done with calipers, ruler, and pen. On the

computer, the calculations are done nearly instantaneously once the program user gives the computer a portrait, an age, and the amount of time elapsed since the picture was taken.

At Louisiana State University (LSU), scientists have developed the Forensic Anthropologic Computer Enhanced System (FACES) to do much the same thing. Using live models of various ages, as well as skulls, they created a database of a number of characteristics, including tissue thickness at various points on the face, the wave time for ultrasound impressions to bounce off a skull, and many others.

As of 1996, the LSU scientists had examined 500 people for their database and had used the database to do some age-progression calculations for missing youngsters, based on the facial characteristics of their parents, as well as old pictures of the children. This enhancement of Barrows and Sadler's idea— that not everyone ages alike, but that people in the same family age in similar ways—has led to some remarkable likenesses. When LSU age-enhanced images are shown on the popular "help us find these people" types of police-dramatization television shows, telephones of police agencies where the disappearances occurred ring off the hook, often years after the disappearance.

At Colorado State University at Fort Collins, scientists are using computer-aided design (CAD) technology, as well as lasers, to scan skulls at 50,000 points, and then to create a three-dimensional (3D) surface mesh. As the laser bounces against the skull, sensors determine the time it takes for the strike and rebound. A photo is then scanned into the computer, and the photo is mapped over the 3D mesh on the computer, to see whether the skull would match the face in the photo.

Facial-reconstruction and age-progression techniques also led to finding a murder suspect missing for years.

On December 7, 1971, police discovered the bodies of five people in a 19-room Victorian mansion in Westfield, New Jersey.

Alma List, 85; Helen List, 45: Patricia List, 16; John List, Jr., 15; and Frederick List, 13, had been shot in the head about a month earlier.

Police found a .32-caliber revolver and a 9-millimeter pistol that had been used in the killings. They also found a five-page confession by John List, a mild-mannered, church-going insurance salesman and accountant who had called Westfield High School on November 9 and said that he was removing his three children from school for an extended trip. After the police found the bodies of List's wife, mother, and three children, they also found his car, parked at Kennedy Airport. List became one of the FBI's most sought fugitives, and the Westfield police chief, James F. Moran, continued to follow leads in the case even after his retirement in 1986.

List apparently moved to Denver in early 1972, assumed a new name, went to work as an accountant, and joined a Lutheran church. He was remarried in 1985 to a woman he met at a church social and he moved to the Richmond, Virginia, area in 1988. He was arrested there on June 2, 1989.

In early 1989, Frank Marranca, the head of the Union County, New Jersey, prosecutor's homicide division, who had kept the List case open for years, asked the producers of a television show, "America's Most Wanted," to do a segment on the List case.

Each Sunday night, beginning in February 1988, "America's Most Wanted" has re-created three, usually brutal, crimes and asked its approximately 22 million viewers, who watch on some 125 stations affiliated with the Fox television network or independent stations who buy the show from Fox, to call a toll-free hot line to report information about any of the cases depicted. Another show, "Unsolved Mysteries," also focuses on fugitives in unsolved criminal cases.

Margaret Roberts, the managing editor of "America's Most Wanted," decided to conduct an experiment. She asked Frank Bender, a Philadelphia sculptor who does forensic reconstructions, to make a bust of List, using 20-year-old photographs and descriptions to reconstruct what he might look like at age 63. "Our show hinges on the indelible image of the human face," Roberts told *The New York Times* in describing the experiment.

The bust, which showed the bespectacled List with wrinkles and graying, receding hair, was shown on the May 21 episode of "America's Most Wanted" and drew immediate response. Within days, 300 telephone tips were received, 200 of which federal officials considered worthy of checking out. One led FBI agents from the Richmond office to the home of Mr. and Mrs. Robert Clark. Mrs. Clark was incredulous when FBI agents told their story but was stunned by the resemblance of her husband to both the old photographs of List and the new photographs of the bust.

FBI agents confronted Clark in his Richmond accounting office; he denied he was List, but his fingerprints matched those of the New Jersey fugitive, and he also had a scar from a mastoid operation behind his ear. He was extradited to New Jersey and later convicted on five counts of murder.

Bender is one of the three founders of the Vidocq Society, named after the founder of the French *Sureté*. The society, based in Philadelphia, is a group of more than 80 mostly retired people from all walks of the investigative community—from laboratory personnel to police officers, FBI agents, and attorneys—who meet monthly to study one "cold case." They often invite the investigating officer for the case, and over a lengthy lunch, they discuss possible angles to follow up or suggestions for exploring new investigative avenues. Since its founding in 1990, the society has explored more than 50 cases, and some long-cold cases have been solved, thanks to the society's help.

The subject of murder has so captured the national attention that at the blue-blooded Amherst College in Massachusetts, Austin Sarat, the William Nelson Cromwell Professor of Jurisprudence and Political Science, began teaching a course in 1995 called, simply, "Murder." With a reading list ranging from *Macbeth* to *In Cold Blood* to *Eichmann in Jerusalem,* the course attracted 300 students, nearly 20 percent of the school's enrollment.

Sarat told *The New York Times,* "I tell my students, 'You think you understand O. J., but you don't. If you want to understand O. J., come with me on a strange journey. You've got to read Sophocles's *Oedipus Rex* and *Macbeth* and *Crime and Punishment* by Dostoyevsky.'" Sarat believes Americans are so fascinated by murder because "it is the one experience of our lives that we can't experience."

Forensic scientists are not above performing outlandish experiments in their efforts to determine identity. Perhaps the most outrageous of these was recounted in the Crafts trial by Harold Wayne Carver, the Connecticut state medical examiner, who pronounced Helle Crafts dead in March 1987.

The experiment Carver described turned a number of stomachs; he had asked state police forensic-lab personnel to put a dead pig through a wood chipper to see whether the fractures in the bone resembled those from the pile of human compost found in Southbury. A pig was used because pigs have skin and hairs similar to those of humans. The results of that experiment were countless small bone fragments similar to those found on the riverbank, as well as a number of larger pieces that contained significant amounts of tissue attached to them. Carver speculated that the larger pieces of Helle Crafts had been taken away from the crime scene by scavenging animals.

Despite this determination, after more than a week of deliberations, a single juror held out against convicting Richard Crafts and a mistrial was declared. On November 21, 1989, however, after a 10-week second trial, at which most of the forensic evidence was again presented, Richard Crafts was convicted of murder.

3

INSIDE THE LAB:
CHEMISTRY +
PROBABILITY = CRIME

The week before H. Wayne Carver, Connecticut's chief state medical examiner, shocked the jurors and audience at the so-called "wood-chipper trial" of Richard Crafts in New London, the prosecution had offered detailed and damning evidence about hair and fibers that linked Mrs. Crafts's body to her hairbrush, apparent body parts found by the banks of the Housatonic River, and a chain saw owned by her husband.

Harold A. Deadman, a special agent in the microscopic analysis unit of the FBI Laboratory and a world-renowned expert in hair and fiber analysis, told the jury that strands of hair taken from Helle Crafts's hairbrush and sponge rollers matched hair strands found among wood chips and human remains discovered along the Housatonic River in Southbury in January, 1987.

Deadman had examined 19 hairs from the hairbrush and rollers, 15 of which had similar characteristics; he also found that 15 hairs of those taken from the river's edge had similar characteristics. Nine of the 15 hairs from the river had the same microscopic characteristics as the 15 from the brush and rollers, which all had the same microscopic characteristics.

A Connecticut State Police criminologist testified that she had taken samples of hair from the three Crafts children and the babysitter, which did not match the strands Deadman used for comparative purposes, although some other strands from the hairbrush did match hair from at least one of the Crafts children.

Deadman testified that the hairs from the riverbank that matched hairs from the brush and rollers all had "very distinctive and unusual characteristics," including "shouldering"—a protrusion from the hair shaft that is extremely rare. This led him to believe that there was "very strong evidence that the hairs came from the same source as the hairs taken from the brush and rollers."

The next day, John Reffner, a forensic fiber analyst and consultant to the state police laboratory, told the jury that blue-green cotton material found among the wood chips and remains on the bank of the river matched material removed from the chain and handle of Crafts's chain saw. The chain saw, recovered from the river by state police divers on January 8, 1987, had been purchased by Richard Crafts at Tom's Saw Service in Georgetown, Connecticut, in January 1981, according to a sales receipt.

Henry Lee, testifying—at both of Crafts's trials—as the final state forensic witness and making his patented broad reconstruction of the crime, later told the jury that the pieces of cotton material matched a larger piece that was from the "neck area of a T-shirt," and that had wood chips, hair, vegetation, and blood on it.

Hair and fiber are just two types of *trace evidence,* which also includes paint, glass, dust, and just about anything that may be taken from or left at the scene of a crime by a victim or perpetrator. According to the exchange principle enumerated in 1928 by Edmond Locard, whenever two objects come into contact, there is a transference of material from one to the other.

Fiber Analysis

The FBI has conducted hair and fiber analysis since the 1940s, and each year its examiners consult on more than 40,000 cases for local law-enforcement agencies. Because of the small size of trace evidence, the difficulty of establishing individualization and comparison, and the need to establish statistical probabilities that the evidence in question actually was in contact with a victim or an accused, this evidence often is not collected and, even when it is, it is often either not used at all or not used properly in court.

There are four varieties of fibers: animal, vegetable, mineral, and synthetic. The most common animal fibers include hairs from sheep (wool), cashmere from the Kashmir goat, and silk fibers (filaments) from silkworms. Vegetable fibers include the most common, cotton, as well as jute and sisal, used mostly in industry. Asbestos is the only mineral fiber commonly analyzed. Synthetic fibers represent about 75 percent of all the textile fibers used in the United States and, consequently, are the fibers most often dealt with in crime labs. Cotton represents about 24 percent of all textile fibers produced in the United States.

The Federal Trade Commission has 21 generic classifications of synthetic textile fiber; the 6 most common classifications are acetate, rayon, nylon, acrylic, polyester, and olefin. Even among these 6 classifications, however, there are more than 1,000 different fiber types. A *fiber type* is defined as having a particular chemical composition that has been manufactured into a particular shape and size, contains particular additives, and has been processed in a particular way. When color variations are added, there are thousands of synthetic fibers that a crime-lab analyst could encounter.

Despite the huge number of fiber-type possibilities, individ-ualizing fibers is easier than individualizing hair, precisely because each fiber has been in some way "manufactured," while hair occurs naturally. It is relatively easy to tell human hairs from other animal hairs and also to differentiate between scalp hairs and hairs from the pubic region, arms, legs, eyebrows, or beards. There also are racial characteristics of hair.

Once these things have been determined, however, without such an obvious rarity as the "shouldering" that Deadman saw in the hairs from the Housatonic River—a characteristic that can be determined through simple microscopic evaluation—it is diffi-cult to positively match a known hair to an unknown hair. Physical features that often are determined in a detailed micro-scopic and instrumental hair analysis—such as refractive index (how much light passes through), density, and weight—generally are not distinctive enough to make identifications.

Experts admit that in an infinite universe, it would be nearly impossible to say that hairs from one sample are the same as hairs from another sample. What makes identification through hair analysis possible within reasonable statistical probability is that the universe of possible matches usually is small. For example, it can be assumed that the hairs on Helle Crafts's hairbrush were either hers or those of one of her children (probably a daugh-ter). After hairs from her children are examined and matched with some hairs on the hairbrush, and the other hairbrush hairs are matched with hairs taken from the crime scene, a logical con-clusion can be made that the hairs from the crime scene and the hairs from the hairbrush both belonged to Mrs. Crafts. This is the essential equation of physical evidence. Diagramatically, it might look something like Figure 3.1.

With synthetic fibers, the investigator must look at three spe-cific sets of effects: (1) the composition, such as color, size, cross-sectional shape, and surface; (2) the manufacturing conditions and dye formulation; and (3) environmental and handling effects,

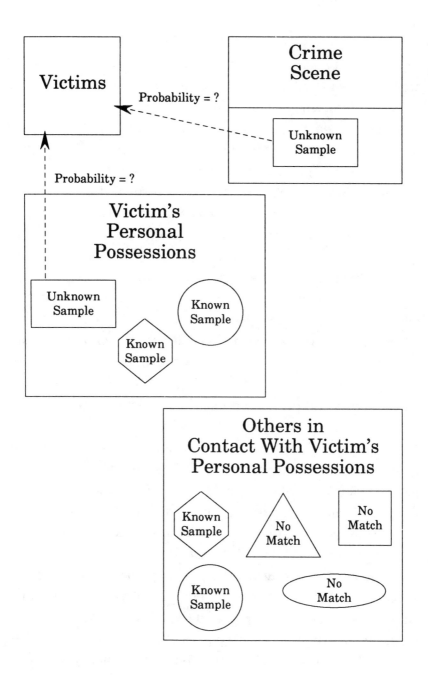

FIGURE 3.1
Logic of Physical Evidence

such as fading and wear. Microscopic examination is done using a *comparison microscope,* which has two plates for samples that are being compared, and a polarized-light microscope and a fluorescence microscope, both of which allow the examiner to look at the fiber under conditions that negate the effect of dye, so that true characteristics can be determined. Polarized-light microscopes act like polarized sunglasses, removing certain light waves from the spectrum that passes through the lens to the viewer's eye. Fluorescent microscopes remove other light waves. In England, the Central Research Establishment of the Home Office Forensic Science Service has created a database of more than 13,000 fibers and their frequency of appearance. The more uncommon a fiber, the greater the impact on a case of matching such fibers.

The case of Wayne Williams, convicted of murder in 1982 and suspected in the Atlanta murders of 12 black teenagers and young men from 1979 to 1981, points out the power of detailed fiber analysis and proper presentation of that evidence in court. As Harold Deadman wrote in his analysis of the case in the *FBI Law Enforcement Bulletin,* "Fiber evidence is often used to corroborate other evidence in a case—it is used to support other testimony and validate other evidence. This was not the situation in the Williams trial (where) other evidence and other aspects of the trial were important but were used to support and complement the fiber evidence."

In July 1979, a string of killings of black male youths began in Atlanta. By February 1981, there had been six similar killings. Atlanta newspapers then reported that of the six bodies discovered, several different fiber types had been found on two of the bodies. (After these articles were published, six more bodies were found before a suspect was caught. All were found in rivers, either nude or wearing only undershorts, an apparent attempt by the killer to remove fiber evidence from any subsequent victims. This example of learning from mistakes, or taking into account information that is seen in news accounts of one's handiwork, is an important piece of information for investigators working on solving serial murders and is discussed in greater detail in Chapter 6.)

The fibers found by the Georgia State Crime Laboratory on the bodies and clothing of the victims were yellow-green nylon fibers and violet acetate fibers. The yellow-green fibers were very unusual, coarse with a lobed cross-sectional shape. Chemists attending a meeting at a research facility of a large fiber producer who were shown photomicrographs of the fibers believed that they were carpet fibers, but the manufacturer could not be determined.

Around 2 A.M. on May 22, 1981, a surveillance team of four Atlanta police officers and FBI agents stationed under and at the ends of the Jackson Parkway Bridge over the Chattahooche River heard a loud splash and an automobile being driven off the bridge. The car was stopped; Wayne Williams was the driver. Two days later, the nude body of Nathaniel Cater was pulled from the river about 1 mile downstream: he had the same type of yellow-green fibers in his head hair. A search warrant was issued for Williams's apartment and car; a similar carpet was found in Williams's home. Police had a possible, if not certain, source of the carpet fibers on the bodies of the Atlanta murder victims.

The key to making such a piece of circumstantial evidence important at trial is to determine how much of that kind of evidence is present in a particular area. If a witness says that an assailant in a corporate office building was wearing a white shirt, that doesn't get investigators very far, because white shirts are often worn by corporate executives and managers.

Investigators in the Williams case did a painstakingly detailed analysis of the particular fibers found, in an effort to establish a credible statistical probability that the fibers came from Williams's carpet. They found that the fiber in question was manufactured by the Wellman Corporation and sold during the years 1967 through 1974. Similar fibers were manufactured by Wellman for many years, but that particular lobal cross-section was manufactured only for that eight-year period. The carpet was composed of a fiber called Wellman 181B. The Wellman 181B was sold in undyed sections to 12 carpet-yarn spinners, who then sold carpet yarn made with Wellman 181B to carpet manufacturers; a number of both yarn spinners and manufacturers had gone out of business since 1974.

Eventually it was determined that the West Point Pepperell Corporation, of Dalton, Georgia, had manufactured the carpet found in Williams's bedroom. It was called Luxaire, and the color was called English Olive. Luxaire was manufactured from 1970 through 1975, but it was only made using Wellman 181B fiber for one year, between 1970 and 1971. Records indicated that in 1971 and 1972, 16,397 square yards of Luxaire English Olive were sold by West Point Pepperell in its entire Region C, composed of ten southeastern states, including Georgia. (The figures for 1972 were included just in case there was Luxaire English Olive manufactured in 1971, but not shipped until 1972; in fact, only 5,710 yards were sold in 1971 and 10,687 sold in 1972, most of which probably no longer contained Wellman 181B fiber.) This is a minuscule amount of carpet; residential carpeted floor space in 1979 was estimated at 6.7 billion square yards.

Investigators made a couple of assumptions for the purpose of calculating the probability that the fibers on the bodies came from Williams's carpets:

- Sales were the same in every Region C state.
- Luxaire English Olive was installed in one 12- × 15-foot room per home in each home in which it was present.

If this were the case—and these assumptions probably are conservative because some homes probably had more than 20 square yards of Luxaire English Olive, much of the 1972 production probably did not contain Wellman 181B, and some 10-year-old carpet probably had been discarded—one could expect to find only 82 homes in all of Georgia with a room carpeted in Luxaire English Olive. At the time of Williams's arrest, there were 682,995 occupied housing units in metropolitan Atlanta alone. Using this figure and the conservative assumptions made by investigators, the chance of randomly selecting a home in metropolitan Atlanta with a room carpeted in Luxaire English Olive was 1 in 7,792. Even if half of all the Luxaire English Olive manufactured in 1971 and 1972 were purchased in metropolitan Atlanta, the chance of finding another home with the carpet, all other things being equal, would be 1 in 1,559.

In deciding which murders to charge Williams with, prosecutors focused on Jimmy Ray Payne, who had been killed about a month before Nathaniel Cater. One reason was that there was a second fiber connection: Fibers found on Payne's shorts matched those of the carpet in Wayne Williams's 1970 Chevrolet station wagon. By finding the number of cars in which Chevrolet installed this kind of carpeting and finding the number of those cars in the Atlanta metropolitan area, it was determined that the probability of randomly finding a person with those fibers on his underwear were 1 in 3,828.

When calculating the possibility that a random individual would have both these fibers in contact with him or her, the probabilities are multiplied: 1 in 7,792 × 1 in 3, 828 becomes 1 in 29,827,776. When multiplied by the probability of finding a person in contact with one of the other fibrous materials in the case (Figure 3.2), the probability becomes almost astronomical.

Paint Analysis

An item of trace evidence even more common than fiber is paint, which easily can be transferred from a vehicle involved in a hit-and-run accident to the victim or from any painted object onto a tool used to break into that object, such as a transfer from a safe,

NAME OF VICTIM									ADDITIONAL ITEMS FROM WILLIAMS' HOME, AUTOMOBILES OR PERSON
Alfred Evans	X	X	X			X			
Eric Middlebrooks	X		X				X		YELLOW NYLON — FORD TRUNK LINER
Charles Stephens	X	X	X		X				YELLOW NYLON / WHITE POLYESTER / BACKROOM CARPET / FORD TRUNK LINER
Lubie Geter	X	X	X					X	KITCHEN CARPET
Terry Pue	X	X	X						WHITE POLYESTER / BACKROOM CARPET
Patrick Baltazar	X	X	X	X				X	YELLOW NYLON / WHITE POLYESTER / HEAD HAIR / GLOVE JACKET / PIGMENTED POLYPROPYLENE
Joseph Bell	X				X				
Larry Rogers	X	X	X	X				X	YELLOW NYLON — PORCH BEDSPREAD
John Porter	X	X	X	X	X			X	PORCH BEDSPREAD
Jimmy Payne	X	X	X	X	X			X	BLUE THROW RUG
William Barrett	X	X	X	X	X			X	GLOVE
Nathaniel Cater	X	X	X	X					BACKROOM CARPET / YELLOW-GREEN SYNTHETIC

FIGURE 3.2
Fiber Evidence in Murder Case against Wayne Williams. *Courtesy FBI.*

door, or window to a crowbar or tire iron. The FBI's National Automotive Paint File and a similar file maintained by the Home Office forensic lab in England have thousands of car top-coat colors on file. The FBI file contains the color and paint type, the car or truck manufacturer that used the paint and the years in which it was used, each manufacturer's designation for the color, and the models on which each color was used.

A known paint sample can be differentiated in a number of ways: by reaction to solvent, analysis of the elements in the paint pigments, or analysis of the infrared spectra in the color. Two techniques are used to make the elemental analysis: *scanning electron microscopy,* which allows the user to see any particle 1 micron (1 millionth of a meter) or larger in size, and *energy-dispersive X-ray analysis,* which identifies compounds by the energy they give off that can be captured on X-ray film. In addition, Fourier transform infrared microscopy is used to determine where on the color spectrum thin sections of the paint fall. Minuscule paint chips can be examined using these instrumental analysis techniques, and extremely fine distinctions in color and paint compounds can be made. As one forensic-lab investigator says, "If you can see it, I can shoot it with these instruments."

Just as paint from a car is left on the car, victim, or other property involved in a hit-and-run accident and can be traced back to an individual car, Locard's formula says that the car doing the hitting will take something away with it. Henry Lee, Commissioner of Public Safety of the State of Connecticut and formerly the Chief of the Connecticut State Police forensics lab, tells the story of trying to determine who caused the death of a state trooper by the side of the road as he was making a traffic stop. Another motorist who was being stopped believed that the officer was hit by a truck speeding down the righthand lane but couldn't describe the truck.

Trucks were stopped farther up the road and examined by state police. Under ultraviolet illumination, an image of the struck officer's shoulder patch appeared as if stamped onto the front right fender of the truck that had hit him.

FIREARMS ANALYSIS

Some of the sensitive analytical instruments used in paint analysis are also used to investigate residue left from gunshots, bombs, or arson. The presence or absence of gunshot residue is an important clue in cases of suspected suicide or murder with a firearm. The earliest chemical tests used in determining who had shot a weapon involved testing particular spots on the hands—either of a dead person or of people who might have killed that person—for two chemicals (nitrites and nitrates) from unburned powder. These tests also picked up traces of two other chemicals (barium and antimony) that are used in the bullet's primer. The *primer* creates the spark to launch the bullet from the gun. Unfortunately, however, these tests often created results too vague for courtroom evidence.

Neutron-activation analysis provided needed specificity, but at high cost. Since the early 1970s, forensic researchers, especially those at the Bureau of Alcohol, Tobacco and Firearms, have looked for easier and less costly methods. The two methods that have become most widely used are flameless atomic absorption spectrometry (FAAS) and scanning electron microscopy using energy-dispersive X-ray diffraction (SEM-EDX).

In *FAAS*, the palm and back of the hand of a person suspected of having recently fired a gun are swabbed using cotton or tape. Tape gives the examiner the opportunity to look under a conventional microscope for large particles of gunshot residue but is more difficult to process than cotton. Whatever is collected is removed using a chemical solution and then is aspirated into the FAAS instrument, where heat causes the chemical atoms to interact and become excited. The instrument determines how much of various chemicals—especially lead, barium, and antimony— are present in the sample; this is recorded on a paper tape.

With *SEM-EDX,* a sample is taken from the hand and put on a slide, which then is examined. First, using the scanning electron microscope, the examiner searches for particles that look like gunshot residue (i.e., that have the characteristic morphology—or shape—of gunshot residue). While particles that include lead can be present on many people's hands due to their occupation— especially people who work in various automotive repair jobs—a gunshot-residue particle of lead is pockmarked by the explosion of heat and gas. This "suspect particle" is then examined by EDX to determine its chemical composition.

With all this instrumentation only a few steps away, the forensic investigator concerned with the firearms themselves— as well as with the marks made by other tools—continues to use a basic 8× to 20× comparison microscope for the vast majority of his or her work. Every firearm is individual. In addition to the three variables of *rifling* of the barrel—lands, grooves, and twist—every barrel has subtle distinctions. In the process of manufacture, the tools used to make a gun are worn down with each succeeding gun manufactured. Consequently, gun number 1,000 in a production run will have a barrel that is configured slightly differently from the barrel of gun number 100, and this difference will be apparent when a trained eye looks at bullets fired from the two guns.

The firearm examiner takes a known bullet or shell casing and a bullet or casing fired from a suspect gun under laboratory conditions and compares them under the comparison microscope. Obviously, the basic bullet size must match; the *caliber*—the bore diameter of the barrel, measured in hundredths of an inch or millimeters—and other specific characteristics will help to identify the manufacturer of the gun from which the ammunition was fired. This information allows the examiner to say only that an unknown bullet "could have been fired by this gun." Then a closer examination must begin.

Matching a casing taken from a test firing and a casing taken from a crime scene often is easier than matching bullets taken from a test firing and a crime firing, because bullets are changed by their impact with anything that gets in their way. A bullet fired in a laboratory into a tank of water will be affected differently from a bullet that goes through a victim's skull or one that goes through glass and then lodges in wood.

Given that the bullets could have come from the same gun, the investigator examines the bullets under the comparison microscope, embedding each bullet in a wax mounting that allows the bullet to be turned and examined from all angles.

The comparison microscope was designed by Philip Gravelle, a microscopist, in the 1920s. With his knowledge that manufacturing left subtle differences on each firearm manufactured, Gravelle set about to find a way to overcome the fallibility of human memory when looking at a succession of bullets under the microscope and trying to compare them. The comparison microscope was first used in 1927 by Calvin Goddard, a medical doctor who changed his specialty from medicine to ordnance while in the U.S. Army during World War I. After the war, he returned to medical practice at Johns Hopkins University and, in 1927, used the comparison microscope to reexamine the bullets that firearm experts had said absolutely came from the gun of Nicola Sacco, the Italian anarchist convicted of murder in Massachusetts in 1921, along with his friend, Bartolomeo Vanzetti.

Bullets that match should be scarred in the same way—within reason, because every successive bullet fired from a gun will make a lasting change on the barrel—from the way the lands and grooves in the barrel, and the left or right twist of the barrel, affect the bullet.

Tool-Mark Analysis

Examiners also concentrate on the marks that tools make when they come in contact with another item. For instance, if a crowbar pulls a door jamb open, an examiner should be able to match the mark on the door jamb against a particular crowbar. The tools are examined for foreign deposits, while the mark is examined to determine the type and size of the tool used and any unusual identifying features.

Firearm and tool-mark examiners also are becoming more involved in investigating cases of altered identification numbers, especially in automobiles, because of their knowledge of metallurgy. Professional car thieves and so-called "chop shops," which provide auto-body repair shops with replacement parts for new cars taken from stolen cars, often alter the vehicle identification number (VIN) on the engine block and the car's dashboard before reselling it, to make the car less traceable in any theft investigation. The FBI maintains both a VIN Standard File and National Automobile Altered Numbers File, with which it keeps track of altered VINs. VINs and other identifying numbers, such as firearm serial numbers, often are filed away or removed using some kind of acid; firearm and tool-mark investigators continue to research ways to make the old numbers appear, using a variety of chemical and instrumental analysis techniques.

IMPRESSIONS AND IMPRINTS ANALYSIS

In the same way that gun barrels make impressions on bullets and casings and tools make impressions on what they come in contact with, shoes, feet, and cars—in fact, anything that puts pressure on a surface—can leave an impression that can be matched in the future. If the surface is soft, the impression actually will change the structure of the surface; if the surface is hard, the impression will be more of an imprint, a deposit of residue on the surface. In any case, these can be valuable evidence. (Recall the case of the trooper's patch imprinted on the truck's fender.)

Crime-scene investigators photograph impressions or imprints and often make casts of impressions to remove to the laboratory for comparison with any evidence that might be collected in the investigation of the case, often using plaster of paris or dental stone.

DOCUMENT ANALYSIS

A typewritten extortion letter may go to the tool-mark examiner to determine the machine it was typed on; it also may go to the behavioral investigator to help create a psychological profile of the suspect (see Chapter 6). A handwritten extortion note, on the other hand, would go to the questioned-documents examiner. The chemists also get involved because forensic experts can use paper and ink to determine when an item was written, as well as by whom.

One such investigation was conducted by the Secret Service in late 1988 and early 1989 for the House of Representatives sub-committee on Oversight and Investigation, chaired by Congressman John Dingell of Michigan. Since 1987, Dingell's subcommittee had been looking into possible scientific fraud by scientists either working at or funded by the National Institutes of Health (NIH). One of his main targets was David Baltimore, a Nobel Prize–winning biologist at the Whitehead Institute in Cambridge, Massachusetts.

Walter Stewart, a scientist at NIH, had spent much of his time since the mid-1980s examining the published work of other scientists and writing critiques of their work, often accusing them of sloppy science at best and fraud at worst. Scientific fraud had become a hot topic in the 1980s, with revelations that researchers at such prestigious universities as Harvard and Emory had spent years building up file drawers full of fraudulent lab results and publishing reams of those data in the ever-increasing race for tenure and scientific awards. Often, senior scientists were caught up in a web of a younger researcher's fraud, absently signing off on work that would appear under both researchers' names.

Dingell focused his attack against Baltimore on a 1986 article in the journal *Cell,* in which the senior scientist and a number of coauthor researchers from his lab described the molecular activities of certain reagents on a number of cells with different genetic makeups. A very junior researcher accused one of her superiors in the lab, Thereza Imanishi-Kari, with misrepresenting laboratory data in the paper. Some of the data in the lab books— and, thus, in the paper—were fabricated after experiments failed to produce the desired results, the researcher contended.

Thus began an investigation by the Massachusetts Institute of Technology, where Baltimore was a faculty member (he was named President of Rockefeller University in New York in November 1989, a position he has since resigned to return to research); by the NIH; and, finally, by Dingell's subcommittee. In the midst of these inquiries, Imanishi-Kari's lawyer wrote to the Dingell committee that "all of the data presented in the *Cell* paper were assembled contemporaneously with the scientific experiment and placed in the notebooks in a timely fashion." Researchers usually spend time at the end of each day updating their lab notebooks.

The lawyer's statement, Secret Service forensic experts found, was not true. What they found was that many pages of lab books from 1984 and 1986 were written on the same two pads of paper, and that many of the pages dated 1984 were written with ball-point pens that were not even made until 1986.

Secret Service agents used a piece of equipment called an electrostatic detection apparatus to determine the sequence in which the pages from the two pads of paper were written and torn off. In one instance, page 41 from the 1984 notebook was the page on the pad directly above page 113; in another instance, page 25 in the 1986 notebook was written and torn from the same pad from which page 30 from the 1984 lab book was subsequently written and torn.

The Secret Service forensic services division's chief document examiner, John Hargett, explained to *Science* magazine how the electrostatic detection apparatus works: "You place a document on top of a metal mesh and stretch a material like Saran Wrap tight across the top. It sucks the document and the plastic tight onto the mesh and uses something like Xerox toner," which clings to the electrostatically charged areas. When the original document is pulled away from a superimposed clear plate, all of the images on the page—those that were written on it and those that were impressed on it from pages written and torn from above on the pad—are apparent. By putting all suspect original pages against each image and matching them, investigators can determine which pages were above which other pages on a pad of paper and even the exact order of the pages. Rewritten words also can be determined.

Hargett also found changed dates in which a different pen was used to try to create another date, rather than just crossing out the old date and putting in the new one, as he suggested the average person would do if he or she merely got the date wrong. (Imanishi-Kari, in fact, admitted to the Dingell committee that she often is a sloppy note keeper and did record some data as much as two years after the experiments were done, but denied that any fraud was intended.)

This kind of work is old hat to the Secret Service. It maintains the International Ink Library, which it assumed in the 1920s from the police in the canton of Zurich, Switzerland. The library contains the chemical composition and other information such as date of formulation on 6,000 types of ink. Although the forensic experts have worked on such notable cases as the fake Howard Hughes will and records that implicated war criminals and murderers, the majority of its work involves convicting those who cheat federal programs such as Medicaid and Social Security.

DRUG ANALYSIS

In 1988, while Dingell and his subcommittee brought forensics into Capitol Hill committee rooms and the world of politics, other forensic scientists made a name for themselves in the world of sports. In September of that year, the Canadian sprinter Ben Johnson was stripped of his Olympic Gold Medal in the 100-meter sprint after he tested positive for anabolic steroids. (A number of other athletes tested positive on Olympic drug tests and were relieved of their medals, and three weight lifters from one country withdrew from competition, presumably because they feared that their own steroid use would be detected.) At first, Johnson denied using steroids; then, for months, he claimed that a prerace drink had been "spiked." In 1989, however, he admitted that he had used steroids for a number of years, as, he said, many athletes he knew also did. Since the late 1980s, a number of other Olympic athletes have been forced to withdraw from the games because of having failed drug tests.

Winners in other sports are routinely tested for performance-enhancing drug use. In the 1988 Tour de France, Pedro Delgado tested negative for steroids, but positive for a drug used to mask steroids. Players in professional team sports in the United States can be tested for drugs if their performance causes suspicion. Professional baseball (the American and National Leagues), the National Football League, the National Basketball Association, and the National Hockey League all have provisions in their player contracts for suspensions and treatment programs for players found to have drug problems. Such well-known and top-performing athletes as major league pitcher Dwight Gooden have been suspended and treated for such problems. Chris Antley, a young jockey, was suspended by the state of New York for 30 days in late 1988 for cocaine use. He underwent treatment and rode the greatest number of winning horses at the Aqueduct Race Track meet in 1989, only to voluntarily give up his license again in late 1989 and undergo treatment a second time. Pitcher Steve Howe underwent five suspensions for relapse into drug abuse before he finally retired voluntarily from baseball.

Just before the Super Bowl in January 1990, a Washington D.C., television station reported that in the previous four seasons, at least three quarterbacks—all white—had tested positive for cocaine and had been allowed to keep playing, while black players who tested positive were routinely suspended.

Drug use to enhance athletic performance has long been a problem for regulators of sports. The International Olympic Committee (IOC) bans over 3,000 drugs, and all medal winners are subject to urine tests. In states where pari-mutuel horse and dog racing is legal, state police forensic technicians often spend a disproportionate amount of their drug-testing resources testing the win, place, and show finishers and any other animal having a suspicious performance.

Of course, drug use, and drug testing in sports only mirror a larger national phenomenon. Evidence in suspected cases of driving under the influence (of alcohol or other drugs) keeps police lab technicians busy, and drug use often is apparent in many crime scenes. The use of certain drugs is criminal in and of itself, and drug abuse goes hand in hand with other crimes against people and property.

It is estimated that 75 percent of prisoners in American suffer from substance abuse (either alcohol or other drugs). In the late 1990s, politicians around the country have been reexamining the tough drug-crime sentencing laws of the late 1970s and early 1980s, asking whether taxpayers should continue to build prison space for an increasing population of nonviolent drug offenders.

Drug-treatment programs in prisons are overwhelmingly successful, with graduates of "therapeutic community" programs, in which prisoners in treatment are separated from the general population, showing markedly lower rates of recidivism after release. Therapeutic-community programs outside prison cost half as much as those inside prison, and a year of residence in a therapeutic community costs taxpayers less than a year in a prison's general population. Increasingly, states are creating "drug courts" or similar programs in which judges can sentence individuals to strict treatment regimes rather than prison.

Drug-use examinations are also part of the forensic evidence in many civil cases, especially in crashes of trains and airplanes, and even in the cases of ships running aground. The captain of the *Exxon Valdez,* which spilled 11 million gallons of oil off the coast of Alaska in early 1989, allegedly was so drunk before he boarded the ship that an untrained and unlicensed mate was asked to steer the ship through difficult waters near Prince William Sound. The issues of whether employers can mandate drug tests and when they can administer such tests have been clogging the federal and state courts since the mid-1980s. Essentially, preemployment drug tests are legal; tests when there is reasonable suspicion that someone is working under the influence are legal; and even random tests of those who work in public safety, public accommodation, and dangerous jobs are legal.

Although the chief use of drug tests has been for professional athletes and employees, other applications have been suggested. In the mid-1990s, the issue of drug tests for students reemerged as a hot topic. School-board members in some districts have pushed for policies mandating drug tests for all student athletes and even for any student wishing to participate in any extracurricular activity. For the most part, these efforts have gone nowhere, running into fierce opposition from civil libertarians and from parents who don't want schools usurping their rights to raise and discipline their own children as they choose. A corollary to this movement by policymakers is the creation of home drug tests being marketed heavily to parents for them to test their own children for drug use. Again, these made their appearance in the early and mid-1990s but don't seem to have caught on among the general public.

Private drug-testing laboratories process well over 1 million urine samples per month in the search for employees or potential employees who use such drugs as marijuana, cocaine, amphetamines, barbiturates, and benzodiazepines (found in tranquilizers such as Valium and Librium, and also in a number of medications taken for high blood pressure).

These laboratories mostly use a two-stage test. In the first stage, the lab mixes the urine sample with a screening assay. The assay contains antibodies for a number of different drug classes, as well as compounds of the drugs themselves, which are tagged with chemical enzymes. In a clean sample, the antibodies and the traces of the compounds themselves bind. In a sample containing metabolites of drugs—the drug by-products after being broken down by the body's metabolism—the metabolites compete with the tagged compounds for antibodies to bind with.

By measuring the absorption of light that passes through the processed sample—a technique known as *spectrophotometry*—

it is possible to see how much, if any, of the enzyme-tagged drug remains, a tip-off that the antibodies have bound to drug metabolites present in the sample.

Samples that appear to be positive on first-round testing often are subjected to a second, more rigorous round of testing, using *chromatography and mass spectrometry,* a process in which a beam of electrons is used to cut—or fragment—the compound, producing ions of different masses. Each compound has its own mass spectrum by which it can be identified after the instrument scans the particular spectrum and consults its computer database. A chromatography and mass spectrometry run, which takes a couple of hours and costs over $100 to administer, can identify drug compounds in as low a concentration as one part per billion.

It was gas chromatography (GC) and mass spectrometry (MS), combined with a new method for preparing samples, that tripped up Ben Johnson in 1988. Although Robert Dugall, the IOC's chief medical officer, said that his laboratory in Montreal had checked Johnson five times since 1984 and had never found stanozolol, the synthetic testosterone hormone found in Johnson's 1988 test, Dugall had never used GC-MS because less-sensitive first-round tests had not tipped him to any possible drug use.

In 1985, however, Dugall and two other researchers, Manfred Donike in Cologne, West Germany, and Don Catlin at the University of California at Los Angeles, working independently, found that preparing samples for GC-MS using one of two different chemicals enhanced detection of the 10 different metabolites produced when the body breaks down stanozolol. Johnson's claim that his drink had been spiked shortly before he ran his world-record 100 meters was refuted by the tests, which showed 2 metabolites present that appear only after long-term use of the steroid.

Laboratory techniques have made anabolic steroids detectable since around 1974; before that, the IOC did not even list steroids as banned drugs. Since the advent of tests able to detect steroid use—and more so since the development of newer, more sophisticated tests—athletes have turned to diuretics in an effort to increase urine production, rid their body of steroids faster, and dilute the steroids voided beyond detection. Athletes have also begun using the drug Probenecid, which lowers the amount of drugs passed in the urine. The IOC also bans use of these drugs.

FORENSIC TOXICOLOGY

Ben Johnson's downfall in Seoul and the mass testing of employees and potential employees is just the latest in a journey lasting more than a century-and-a-half in forensic toxicology. It began in 1840, when Marie Lafarge, a 24-year-old French woman, was accused of poisoning her husband Charles—first with a cake she baked for him to eat during his trip to Paris on business, later with venison, truffles, and other food when he returned home. Charles Lafarge died on January 14, 1840, nine days after eating the first of the apparently arsenic-laced meals.

For centuries, people had looked for ways to distinguish death by deliberate poisoning from death due to improper dosage of medicines or from one of the viral or bacterial illnesses that often caused similar symptoms. It was known that arsenic is tasteless and odorless, and the violent cramps, vomiting, and other symptoms of arsenic poisoning are very similar to those of one of the most prevalent diseases throughout the Middle Ages and early modern times: cholera.

In 1836, James Marsh, a chemist with the Royal British Arsenal in Woolwich, near London, discovered a method for detecting minuscule amounts of arsenic. Marsh used a U-shaped tube, with a nozzle at one end. He put a piece of zinc near the nozzle and placed the fluid sample on the open end. Then he added acid. When the fluid and acid reached the zinc, if arsenic was present, arsenine gas was formed and escaped from the nozzle, where it coated a cold porcelain cover with a black deposit.

Later refinements heated the liquid as it flowed through a tube to produce the arsenine gas.

Four years after Marsh's discovery, a refined apparatus was used by French scientists to convict Madame Lafarge. In effect, Marsh created a primitive gas chromatographic system, except that it searched for only one substance. Throughout the 1800s, scientists would learn infinitely more about chemical compounds and how they behave when either mixed with other chemicals or subjected to heat. Over that century, scientists also would learn to detect other poisons. Killers and scientific sleuths would continue to play a cat-and-mouse game.

Throughout this process, the human element also has loomed large. The first tests at the Lafarge trial, using the Marsh system, yielded no evidence of arsenic; senior French scientists attributed this result to young, untrained examiners using the apparatus.

Although today's instrumentation manufacturers say that there is only a tiny possibility of instrument error being responsible for a false-positive result, critics of mass drug testing argue that the possibilities for human error are great. Studies of all kinds of laboratory processes, from Pap smears to blood tests, routinely point out high error rates—and the consequences of falsely labeling someone a drug user are enormous.

What is more frightening is the number of employers who decide not to spend $100 apiece for tests using chromatography followed by mass spectrometry; instead, they are turning to cheap and easy home-testing kits, are "training" in-house employees to do the test, and are not heeding the warning that these are only preliminary tests. The possibilities for abuse of the testing, for imprudent political decisions, and for just plain sloppy work done by unskilled testers are astronomical under these conditions.

Whether the company does home testing or uses a commercial lab, it will run into a problem maintaining what is called the chain of custody. *Chain of custody* is a law-enforcement term that describes the route a piece of evidence travels from the time it is collected until it is presented in court. If that chain has been broken, or even may have been broken, the evidence's value becomes uncertain, and judges usually warn jurors that it is difficult to find an individual guilty beyond a reasonable doubt based on uncertain evidence. The level of proof required to destroy a person's reputation and to deny that person a job should be no less; unfortunately, that is not so.

Often, in a misguided attempt to avoid evasion in drug testing, the violation of the test-takers' civil liberties, not to mention dignity, goes even further. To thwart the switching or purchasing of urine samples, many drug tests are performed with a supervisor or potential employer watching the production of the sample.

Sometimes the instruments used are so sensitive that they pick up traces of drugs that have been introduced into the system totally by accident. The case of cocaine found in the urine of racehorses in California shows how this might happen. In the fall of 1988, California authorities found traces of cocaine—24 to 33 billionths of a gram—in the urine of two winning horses at the Santa Anita racetrack and in the frozen urine samples of two other horses from the previous summer's racing at Del Mar. Four prominent California trainers were implicated.

Although cocaine frequently is used by human athletes as a recreational drug and not necessarily as a performance enhancer, no one could understand why it was being found in the urine of racehorses, especially in such small quantities. There are far more potent and less detectable painkillers and stimulants that trainers could use.

If the doping of these horses was not intentional, what could have caused it? Trace amounts of cocaine are found in some leg liniments used on horses, but the four trainers involved in the alleged doping incidents all said that they used brands with no cocaine. Speculation was that perhaps stable workers with cocaine on their hands introduced it into the horses' systems when bits were inserted into the horses' mouths or when they rubbed down the horses' legs.

One toxicologist's research suggests that this is a likely scenario. Lee Hearn, chief toxicologist for the Dade County Medical Examiner's office in Florida, was puzzled that police were confiscating large amounts of cash that showed traces of cocaine and automatically assuming it to be drug money. Hearn checked piles of $20 bills from a number of Miami banks by washing the bills and putting the residue through a mass spectrometer. The instrument identified cocaine in all the packets.

Hearn then got 135 bills from 12 different cities and subjected them to the same test. All but 4 bills had at least a trace of cocaine. Some bills, Hearn speculated, are used for snorting cocaine, and others come in contact with people with traces on their hands. Grains of cocaine then get transferred from one bill to others in wallets, cash drawers, and anywhere else. Presumably, traces also can get into people's bloodstreams if they lick their fingers, eat with their hands, or even scratch insect bites. With the high sensitivity of contemporary drug-testing equipment, many people could be suspected of using drugs who do nothing more sinister than lick their index finger before counting out a roll of bills.

Explosives and Fire-Residue Analysis

In another corner of the crime lab, residue from bombs is examined, often by using thin-layer chromatography, gas-liquid chromatography, and high-performance liquid chromatography. These analyses can add an extra dimension to the examination of physical evidence at a bombing scene. The pattern of destruction can help investigators to determine where explosives were placed. For instance, investigators were able to determine in the bombing of a Pan American jetliner over Scotland in 1988 the precise piece of luggage in which the bomb had been placed and how the charge was detonated. Pieces of alarm clocks or pressure-sensitive triggers often are found in bomb debris or in the bodies of bombing victims during autopsy, but it is the chemical analysis of the explosive that often can help investigators to determine with precision what the explosive was and where the explosive was made. These determinations often are clues to who set off the explosive. Various federal agencies, from the FBI to the Central Intelligence Agency (CIA), have detailed intelligence records of which crime or terrorist organizations buy their armaments and explosives from which supplier countries and even from which companies.

In *gas chromatography,* used for analyzing fire residue, all of the chemical compounds in a particular sample are determined. A sample is vaporized and sent down a tube by an inert gas, in which all the compounds in the sample separate. Each compound settles in the tube at a different rate, which is called the *retention time.* A reference library of retention times is consulted to find the compound.

Some compounds would break down completely if they were vaporized. For these, *liquid chromatography* is used, in which the sample is carried down the tube by a liquid. *Thin-layer chromatography* also uses a liquid to transport the sample; the sample is then separated into its various compounds by a layer of finely ground silica gel. Accelerants and explosive chemicals move down the column at different rates than do the other substances in a residue—usually ash, charred wood, plastic, or fiber.

For years, terrorist bombings were thought by Americans to be a phenomenon of the Middle East. Sure, there were occasional political bombings in this country, mostly surrounding the civil rights movement of the 1960s, and against abortion and family-planning clinics in the 1980s. Usually, these bombs destroyed only property, although some deaths and injuries also resulted.

The 1990s, however, have brought the terror of bombing to Americans in three new phenomena.

1. Middle East terrorism reached America's shores in late February, 1993, when the World Trade Center in Manhattan was damaged by a truck bomb that exploded in its underground garage. Five people were killed, and more than 1,000 were injured. A number of Muslim radicals have been convicted for this crime in a series of trials from 1995 to 1997.

2. Domestic political terrorism struck in April, 1995, when a truck bomb sitting outside the federal building in Oklahoma City, Oklahoma, sheared off the side of the building. One hundred and sixty eight people were killed, including children in the building's child-care center, and over 850 people were injured. In the summer of 1997, Timothy McVeigh was convicted of the bombing and sentenced to die.

    Later in 1998, McVeigh's accomplice Terry Nichols was convicted of conspiracy to explode an explosive device and multiple counts of manslaughter. The jury did not believe he knew exactly when the bomb would go off or that he had direct knowledge it would be timed to detonate when people could be killed or injured. The jury also refused to sentence Nichols to death; a federal judge imposed the harshest sentence he could by law: life in prison.

3. Serial bombing, the product of the so-called Unabomber, reached a more fevered level through the early and mid-1990s. In 1996, Theodore Kaczynski, a former mathematics professor, was arrested at his one-room shack in North Dakota and charged with being the Unabomber, so named because his first victims were university personnel and airline executives. In January, 1998, Kaczynski pleaded

guilty to four federal counts of murder and admitted per-petrating all of the Unabomber homicides and assaults. He was sentenced to life in prison. (The Unabomber killings were federal charges because the mail was used to send the bombs.)

All bomb investigations are carried out on two tracks: One is to search for a motive; the second is to search for the bomb's components and any "signature" of the bomber. Police detectives follow the first track, while the crime lab focuses on the second.

Forensic investigators look for six things. First, they look at *blast damage.* The extent of the damage can often tell investigators the type and amount of explosives used. Some high explosives create blasts that cause momentary vacuums, sucking windows in, rather than blowing them out. Gun powder must be packed to create an explosion. Plastic explosives don't need to be packed, but they need a charge, a blasting cap, and a timing device.

Second, *unignited explosives* identified visually or through laboratory instrumentation, such as mass spectrometry and gas chromatography, can sometimes link the type of explosives used to a suspect who might have access to that type. Some explosives are available commercially, such as dynamite, used in construction blasting. Other explosives are available only to the military. Still other explosives are homemade, such as the ammonium-nitrate fertilizer bomb that McVeigh used to blow up the Oklahoma City federal building.

Third, investigators look for any *large bomb components.* Portions of pipe are always found when pipe bombs are set off. In the Unabomber case, investigators linked the blasts, which occurred over 18 years in many regions of the country, because of bomb components meticulously carved of wood.

Fourth, *microscopic components* (such as bits of clocks or watches used as timing devices, batteries, wire, or blasting caps) can be used by investigators. Seen under a microscope, these pieces can be identified by unique configurations, much like the lands and grooves in each bullet.

Fifth, *tool marks,* such as those left on wire by wire cutters, can help link a particular tool to a given bomb or a tool's owner to a bomber.

Sixth and finally, *paper, tape, and wrapping paper* can hold clues. The torn end of tape may match the end of a roll owned by a suspect; the suspect's fingerprints may be on tape or packaging material; even the DNA in a suspect's saliva on an envelope or on the stamps of letter or package bombs can all be helpful in a bombing investigation.

4
FINGERPRINTS: PROOF POSITIVE

When Stanley Schwartz (the former Tufts University professor of dentistry) began his examination of the New Bedford, Massachusetts, skeletal remains in November, 1988, there were three sets. Within a month, two more sets of remains had been found, and Schwartz was still poring over dental charts of hundreds of missing women, trying to match the skeletons' teeth with the sometimes incomplete dental records.

The first body identified was that of Dawn Mendes, the 25-year-old mother of a 5-year old boy; she had been missing since September. While Schwartz's diligence paid off, investigators had a stronger identifying mark for Mendes—her fingerprints. On at least one of her fingers, her fingerprint was intact, and police were able to match it with a complete set they had of her from an earlier arrest.

Fingerprints are unique; they are proof positive of identity. Millions of Americans have their fingerprints on file: The military takes fingerprints of every member, to aid in identification after death, while the intelligence agencies routinely fingerprint employees as an aid in any subsequent espionage or counterespionage investigation that might need to be conducted. Law-enforcement agencies routinely fingerprint everyone who is arrested and charged with a crime. Since the early 1990s, an increasing number of states have fingerprinted welfare recipients, in an effort to curb fraud, despite legal challenges from civil libertarians. Some states also fingerprint teachers and others who work with children.

The FBI maintains a computerized reference library of more than 83 million fingerprint cards representing more than 22 million people (nearly 10% of the American population, many with multiple card entries from separate arrests) in its criminal fingerprint register. Over 30,000 fingerprint cards are added to the collection every day, and FBI employees handle thousands of requests for comparisons each year. Some of this comparison work has been eliminated in recent years by the creation of automated fingerprint identification systems (AFIS), artificial-intelligence computer software that allows investigators who have computerized access to the FBI library to search the collection themselves for a match with any set of prints they may have. An increasing number of states now have computerized fingerprint systems, but the national FBI AFIS has never been made fully operational, as a result of a lack of funding from Congress.

This system, described in more detail in Chapter 8, produces a universe of the best few possible matches, which then can be examined closely by trained human fingerprint analysts. AFIS can match reference fingerprint cards either to fingerprint cards taken at the time of a new arrest or to *latent prints*—fingerprints that occur in the normal course of touching objects—removed from a crime scene.

HISTORY

The examination of fingerprints has been the cornerstone of identification for over a century. The discovery that each person's fingerprints are unique was, as is often the case in science, serendipitous. As also is often the case, however, the discovery and use of fingerprints was born of necessity, too; a huge increase in the number of criminals and the difficulty of remembering them led to a simple thought that there must be a systematic way to physically classify people. At first, this classification was done by physical measurements, or *anthropometrics.*

In the 1870s, the files of the French Sureté were filled to overflowing. Files listed each criminal by name, aliases, places he or she frequented, crimes, sentences, and a detailed physical description. Often there also was a photograph. Time changes people, however, and younger officers did not know older criminals, and the files were centralized.

In 1879, a new clerk was hired into the army that maintained this morass; Alphonse Bertillon, a 25-year old son of the distinguished physician and anthropologist, Louis Adolph Bertillon. A sullen, asocial young man, Bertillon nevertheless was curious and well read in the natural sciences. He had read Darwin, had heard of Pasteur, and was at least vaguely familiar with the work of other scientists. He had watched his father and his grandfather, the naturalist and mathematician Achille Guillard, measuring the skulls of people of different races in a primitive search for a relationship between head shape and intelligence.

A few years earlier, Cesare Lambroso, the Italian psychiatrist, had written his pamphlet on "The Delinquent Man," in which he outlined his theory of a primitive—or atavistic—natural criminal. He described this person as having specific physical characteristics, including a particular head size and shape. In addition, Adolphe Quetelet, a Belgian astronomer and statistician, had written that human physical development was subject to natural laws.

Bertillon was frustrated at spending his days scribbling notations on identification cards that referred to people as of "medium" height and having "fair" skin. By comparing the grainy photographs, he saw differences in the size and shape of various body parts. Soon, he was given permission to measure convicts brought to Sureté headquarters for registration; he measured height, length of head, circumference of head, and length of arms, fingers, and feet. Eventually, he took 11 measurements in all and determined that the statistical probability of two people having one measurement in common was 1 in 16, while the chance of two people having all 11 measurements in common was 1 in 286,435,456 (which essentially means that in the United States in 1998, no two people are likely to have the exact same 11 measurements).

The method of physical identification known as *Bertillonage* had been invented. It eventually would become accepted throughout much of Europe and even in the United States in a historical sense; however, his work was a dead end. A far more fruitful approach had been taking shape in India for 20 years.

Beginning in 1858, William Herschel, a British administrative officer in the Indian district capital of Hooghly, had been asking any person who registered an official document to sign it with his or her "sign-manual," an imprint of inked fingers and palm. He originally had thought that this might hold some mystery for people who did not write and who, he had come to find out, often conveniently forgot about contracts. Over time, however, he began to notice that no two marks looked alike, and all were filled with curiously lined patterns—and they never changed. A person could sign a contract in 1858 and another in 1878, and his or her sign-manual would be the same. Herschel did not know that in Japan, sealing a contract with the imprint of a finger or thumb in red or black ink had been done for thousands of years.

Henry Faulds, a British surgeon who was teaching Japanese medical students, did not know that either. Nonetheless, he was curious about finger marks that potters left in clay and in the 1860s, he began a systematic study of what Herschel had called the papillary lines in fingerprints, in a letter Herschel had written to the British journal *Nature.* Faulds's research focused on racial characteristics or inherited traits, both of which turned out to be blind alleys, but his curiosity was piqued.

One day, he heard about a thief who had been caught after climbing over a wall. When he examined the wall, he found fingerprints. Taking inked impressions from the captured man, he compared them to the marks on the wall and found that they matched. Soon after, he encountered a similar case.

He wrote his own letter to *Nature,* which printed it in the issue of October 28, 1880: "When bloody finger marks or impressions on clay, glass, etc., exist, they may led to the scientific identification of criminals. Already I have had experience in two such cases. . . . Other cases might occur in medico-legal investigations, as when the hands of some mutilated victim were found."

In 1888, Bertillon was made chief of the identification division of the Sureté, and the British government sent Francis Galton, a 66-year-old scientific dilettante and the country's foremost expert on *anthropometry* (the study of physical measurements), to meet him. Galton, remembering the letters to *Nature* by Faulds and by Herschel, took the opportunity to make a careful and complete study of identification. Herschel even visited Galton at his laboratory at the South Kensington Museum and taught him how to take fingerprints.

Galton became convinced that fingerprints would be more useful than Bertillonage and set out to do a detailed study. In 1892, Galton's book *Fingerprints* described his proposal for a classification system based on the small triangles that appeared within most fingerprints; he called them "deltas" after the Greek letter and said that any one fingerprint could be classified as having no delta, a delta on the left (of the finger's centerline), a delta on the right, or multiple deltas. The more of a person's fingers were printed, the more distinguishing the fingerprints would be because there are subtle differences among the fingers.

Two years later, the U.S. humorist and novelist Mark Twain made the notion of fingerprinting famous in his book *The Tragedy of Pudd'nhead Wilson,* about a country lawyer of the same name who had the unusual hobby of asking friends to leave their finger marks on a glass slide, which he meticulously filed away. While he was roundly ridiculed for this pastime—it would be a decade after the story was published until England first began to use fingerprints for identification and many more years before the technique was widely accepted by U.S. courts—Wilson used his hobby in a court case to convince a jury that the bloody fingerprints left on a murder weapon were far different from those of two friends, who stood accused of murdering a judge.

The New York City police department began fingerprinting people who were arrested in the city in 1903, and the federal Bureau of Prisons also used the technique in the early part of the twentieth century. At the federal prison in Leavenworth, Kansas, shortly after the turn of the century, a man was brought in who gave his name as William West. Although he said that he had never before been a prisoner there, guards were sure they knew the man. They searched the prison records for William West, and when the records arrived, the William West just admitted bore an uncanny resemblance to the figure in the picture in the prison records, and his vital measurements were similar. Upon fingerprinting the new arrival, however, the registration officer declared that the prints were totally different. The other William West was then brought to the registration area and, although the two men could have been identical twins, they had two distinct sets of fingerprints—as, in fact, do identical twins.

CLASSIFICATION

Over the next half century, the distinctions of fingerprints became ever finer; a methodology for recording the information also was developed. Today, the FBI classifies fingerprints in one of three categories—arch, loop, and whorl—each with subcategories, as shown in Figure 4.1.

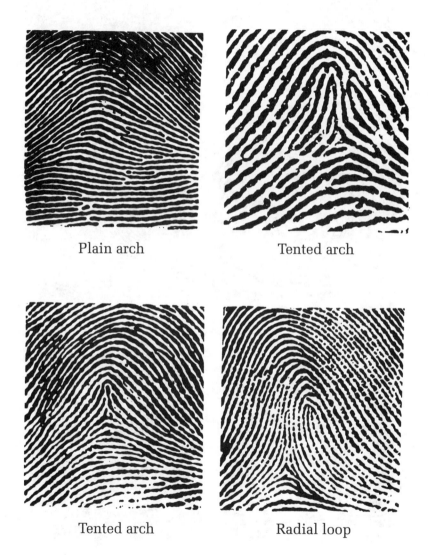

Plain arch   Tented arch

Tented arch   Radial loop

FIGURE 4.1
FBI Fingerprint Classifications

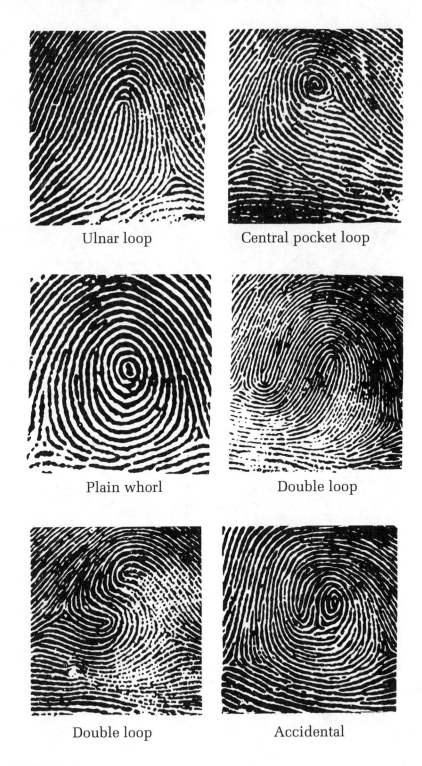

Ulnar loop

Central pocket loop

Plain whorl

Double loop

Double loop

Accidental

FIGURE 4.1
(continued)

Each fingerprint has a focal point, used by the analyst for classification purposes. One type of focal point is the *core,* from which the analyst looks for points where the ridges either *bifurcate* (divide into two after being one) or *diverge* (spear apart after running parallel). The *delta* is the other type of focal point. In Figure 4.2, the cores and deltas are marked with the letters C and D.

Fingerprints are further distinguished by the *ridge count,* the number of ridges that either cross or touch an imaginary line that runs from the core to the delta. FBI technicians put a glass sheet with a red line over the print they are examining, then they move the line so that it runs from the core to the delta, to ensure ridge-count accuracy.

When charting the fingerprints, the analyst uses the letters A (arch), T (tented arch), R (radial loop), U (ulnar loop), or W (whorl) and the number of the finger it refers to, from 1 to 10. When comparing fingerprints, it takes 10 or more points of concurrence to declare the prints a match. Also, thanks to Henry Battley, examiners can match just one fingerprint. In 1930, Battley, working in Scotland Yard, developed a system for filing individual fingerprints, as well as complete sets, and for cross-referencing them. Figure 4.3 shows a complete set of inked fingerprints that have been properly taken.

Matching one set of inked prints to another is a painstaking process—a process made less difficult with AFIS. The uses of inked prints clearly are limited; mostly, such prints permit investigators to determine whether the person who has been fingerprinted—usually as part of the arrest process—has fingerprints on file under different names, or they allow identification of a dead person.

By the turn of the century, fingerprinting with ink will be a thing of the past in many police departments around the country. Already, hundreds of police departments are using "live scan" technology to take full sets of prints: Instead of rolling inked fingers on a pad, clean fingers are rolled over a glass panel

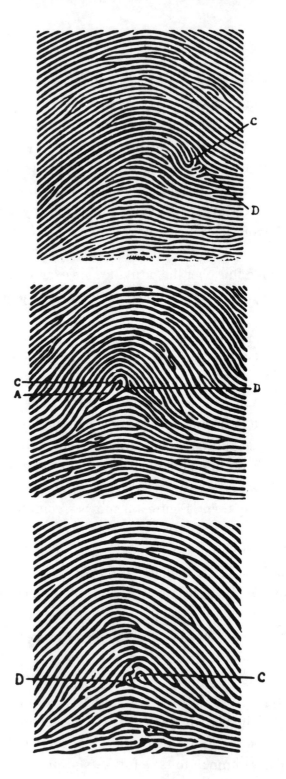

FIGURE 4.2
Cores and Deltas
in Fingerprints

84

set at 30 degrees of angle, to gain a full picture of the pattern. The print can be immediately seen on a computer screen to make sure the transfer is complete. The computer then digitizes each finger's image, translating each pattern or ridge, arch, or whorl into a digital code. Each set of fingerprints uses between 1.5 and 2.5 million bits of information in the computer's memory.

For identification and matching, the computer no longer compares the patterns, but rather the digital code. Nonetheless, it still has to have something to compare the new set of digitized prints

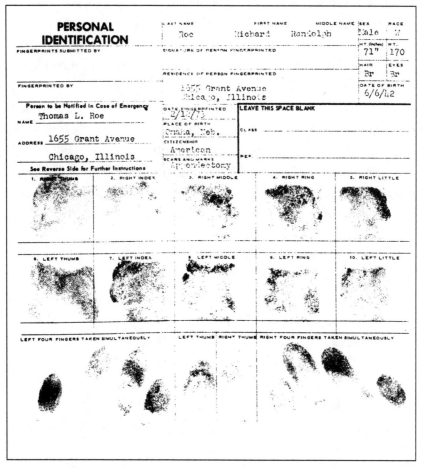

FIGURE 4.3
Complete Set of Fingerprints

to. Since 1990, the FBI has been saying that it intends to transfer its millions of fingerprint files from cards to computers, digitizing the information by the turn of the twenty-first century.
However, lack of funding from Congress has put that project
years behind.

LATENT FINGERPRINT DEVELOPMENT

For solving most crimes, it is necessary to compare latent fingerprints to the reference files of fingerprint cards. Because latent
prints are not taken under ideal circumstances, but are lifted by
a latent-print technician from whatever surface they happen to
be on, they are not always of good quality. Matching latent prints
to fingerprints on file obviously is easier if there are more latent
prints, from more different fingers.

The fingerprints that Faulds saw on the wall in Japan were
the marks of dirty hands, but even the cleanest hands will leave
fingerprints, though they may be invisible to the naked eye. To
lift those prints off the surface, the fingerprint technician must
use one of a number of techniques that have been developed over
the years; most of these techniques are chemical, but the most
recent of these involve laser light.

Fingerprints can remain intact on a surface for years. For
instance, in May 1982, the FBI Identification Division was asked
to examine a 40-year-old postcard in hope of proving that it had
been written by an alleged Nazi collaborator.

Valerian Trifa was born on June 28, 1914, in Campeni,
Romania. He came to the United States in the huge wave of
Eastern European immigration after World War II, arriving on
July 17, 1950. A priest in the Romanian Orthodox Church, Trifa
was consecrated a bishop two years later. In 1957, he became a
naturalized U.S. citizen.

In 1975, however, the Justice Department instituted deportation proceedings against Trifa, alleging that he had misrepre-

sented himself on his citizenship papers in 1957. Trifa was one of hundreds of naturalized U.S. citizens who were accused by the Justice Department of having misrepresented their activities during World War II by omitting their ties to fascist movements throughout Europe. The Justice Department had evidence that Trifa had been a high-ranking member of the Romanian Iron Guard, the violent fascist and anti-Semitic movement that was responsible for the deaths of thousands of Jews in Romania and the assassinations of numerous Romanian political figures in the late 1930s and early 1940s.

The government of West Germany supplied the U.S. Justice Department with a number of documents pertaining to Trifa, some of which the FBI examined for latent prints; one was a postcard allegedly written by Trifa and addressed to Heinrich Himmler, head of the Nazi SS, and dated June 14, 1942. The West German government insisted that the documents not be damaged in any way by the examination, so the FBI used one of its newest and most sophisticated techniques, laser examination. A latent impression of a left thumbprint was developed, and FBI fingerprint analysts matched 11 reference points on the print with an inked print of Trifa. This was among the most powerful evidence presented at Trifa's deportation hearing, and on August 13, 1984, Trifa was indeed deported to Portugal.

The next section describes older, more conventional latent-print development techniques. The section following it describes laser technology. Finally, a new photographic technique is discussed.

Conventional Techniques

Dusting the crime scene may be a familiar phrase to readers of detective fiction and police procedural books and to watchers of police and crime shows on television. This is the commonest, simplest, and oldest latent-print developing technique, in which a powder—usually a gray or black powder—is dusted on wood, glass, metal, ceramic, or other surfaces in the hope that fingerprints will become apparent. The powder, which is made of resinous polymers that can be mixed in literally hundreds of ways, adheres to the skin oils that invariably are left behind on anything touched directly with the fingers. Gray powder is used on dark surfaces or mirrors, while black powder is used on lighter surfaces, in an effort to maximize photographic contrast. Before the technician actually lifts the fingerprint off the surface with tape to which the powder adheres, he or she photographs the print with a special fingerprint camera.

In recent years, those concerned with health have created fingerprint powders made from organic substances, in an effort to limit technicians' long-term exposure to such dangerous inorganic components as lead, cadmium, copper, silicon, and mercury, which were present in most commercial fingerprint powders in the past. In addition, fluorescent and phosphorescent fingerprint powders have been created to solve the contrast problem of developing fingerprints on many colored surfaces.

For fingerprints on paper, cardboard, unpainted wood, and other porous surfaces, fingerprint technicians often use chemical fuming techniques. *Iodine fuming* has been used for more than half a century. Iodine crystals vaporize rapidly when subjected to heat and produce violet fumes that are absorbed by skin oils. When a sample is put in an iodine fuming cabinet or "shot" with an iodine fume gun, any latent prints will absorb the iodine fumes and become visible, appearing yellowish-brown against the background. The prints are visible only as long as the fumes last, so they must be photographed immediately. In addition, old prints often don't develop well using iodine, and the vapors are toxic to the technician and corrosive to other materials.

Another form of chemical fuming is *cyanoacrylate fuming*, discovered in 1982 by latent-fingerprint examiners at the U.S. Army Criminal Investigation Laboratory in Japan and at the Bureau of Alcohol, Tobacco and Firearms. This method is especially useful in developing latent prints on household items such as plastic bags, plasticized electrical tapes, Styrofoam, aluminum foil, cellophane, and rubber bands. Cyanoacrylate is the chemical used in Superglue. The glue vapors adhere to the friction-ridge residue of the latent prints, harden, and build up the ridge detail as more particles condense. The sample must be glue-fumed in a compartment that is as airtight as possible.

The chemical *ninhydrin* (triketohydrindene hydrate), either in solution or as a white powder, has been used to develop latent prints since 1954. The traces of amino acids present in perspiration bind with the ninhydrin, and the prints often begin to appear within one hour, although prints may continue to become visible a day later. Heat speeds up the process, and latent-print technicians often can be seen literally ironing their ninhydrin-treated samples. Ninhydrin is the most common chemical reagent used for developing latent prints on paper.

Silver nitrate reacts with the sodium chloride (salt) in perspiration to develop latent prints in much the same way that ninhydrin reacts with amino acids. In a systematic approach to examining samples for latent prints, silver nitrate examination is done after chemical fuming and ninhydrin examination, because the silver nitrates solution washes away traces of oil and amino acids from the paper being examined.

Gentian violet is the last of the commonly used latent-print development techniques; gentian violet or crystal violet is used to stain nonliving epidermal cells or perspiration that has been left on almost any type of surface. Tape is used to remove whatever might be present from the surface to be sampled, then it is run through the solution of gentian violet. Excess dye is removed with plain cold tap water.

Laser Techniques

Laser technology, first used by the FBI in 1978, has done nothing short of revolutionizing the development of latent fingerprints.

The methodology of using the laser for latent-print developing is rather simple, as shown in Figure 4.4. The FBI's equipment is shown in Figure 4.5.

There is no need to pretreat the sample, and the laser often is used before any other methods. However, because other methods may enhance the ability to develop prints, another pass with the laser often is made between uses of each of the other techniques. Fluorescent fingerprint powders have proved to be especially helpful in enhancing images for the laser examination. Lasers can also be used to get workable latent prints off surfaces that never surrendered prints before, including the flesh of human corpses.

As shown in Figure 4.4, the argon laser output is turned on, and the sample is passed under the viewing area. The examiner looks at the sample through the protective filtering lens; large

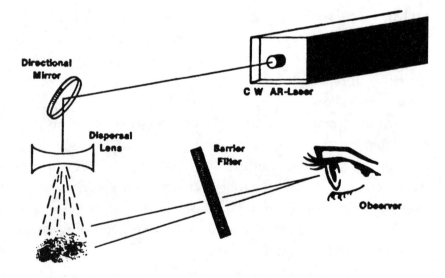

FIGURE 4.4
How Laser Detects Latent Fingerprints

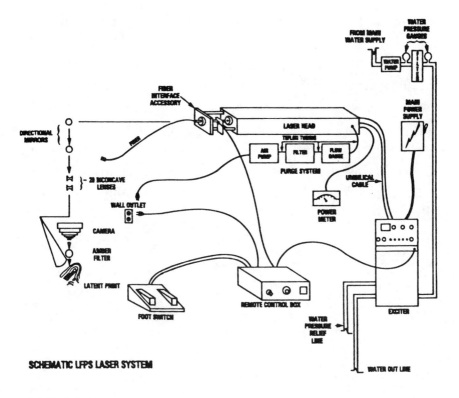

SCHEMATIC LFPS LASER SYSTEM

FIGURE 4.5
FBI Laser System for Detecting Latent Fingerprints

samples can be looked at through a fiber-optic cable, allowing the examiner to focus on particular areas. A polarizing filter can also be used with the fingerprint camera.

Direct Photography

In 1997, Detective John Brunetti of the West Haven, Connecticut, police force came up with a novel way of getting workable latent prints off plastic bags. For years, he had used glue fumes but found that often, the fumes clung to parts of the bag other than the fingerprints, making it difficult to pick out individual prints. Also, once chemical fumes are used, no other latent-print removal technique can be used. Instead of using those techniques,

Brunetti put the plastic bag into a photographic enlarger and filtered light directly through the bag. He has found that with this technique, he can photograph individual fingerprints with a minimum of distortion. Brunetti first tried the technique in a drug case where he was trying to get satisfactory photographs of fingerprints from 216 plastic bags containing crack cocaine.

Each day, latent-print technicians throughout the country and the world dust, fume, laser, and photograph millions of samples in the hope of identifying victims, perpetrators, and witnesses of crime.

5

DNA TYPING:
A GENETIC FINGERPRINT?

In the summer of 1987, a Florida prosecutor about to try a rape case became intrigued by a story about a British manhunt in which police had collected blood samples from over 1,000 men in an attempt to math the DNA they obtained from these samples to the DNA in semen samples from two brutal rape–murders. He decided to test the U.S. legal system's ability to digest new scientific information. Using a technique less than a decade old, scientists at a New York laboratory compared samples of blood from the suspect, Tommie Lee Andrews, a 24-year-old Orlando pharmaceutical company employee, and semen samples from vaginal swabs taken from the victim.

The samples matched, and Andrews became the first person in this country convicted of a crime based on the uniqueness of his DNA. Within the first decade of its use, forensic DNA typing has been hailed by some as the greatest crime-solving tool since fingerprinting. Nearly everyone believes that in theory, it is a powerful tool; however, throughout the 1990s, an argument has raged about its actual practice:

- Are the tests performed by police labs and private labs conducted with the most rigorously accurate scientific practices?

- Is the population genetics on which DNA examiners testify that they are as certain as can be that the defendant is guilty flawed?

- Can a DNA test that shows the defendant to be not guilty be disregarded by a jury, in favor of a statement from a compelling victim that the defendant raped her?

These kinds of issues have been thrashed out in state and federal courts, at the trial and appellate level, since 1991.

DNA typing quickly acquired the title of "genetic fingerprinting." In fact, DNA Fingerprint is the trademarked name given the process by Cellmark Diagnostics, a Maryland company that licensed the technique used in Great Britain; Lifecodes, the New York company that also does commercial DNA identification, has trademarked its technique as the DNA-Print test.

Forensic scientists see genetic fingerprinting as a far more powerful tool than conventional serology, in which blood, semen, saliva, and other body fluids can be tested for the presence of certain blood-group factors, such as the simplest ABO, or Rh-negative/positive typing, or protein markers. Police and prosecutors often tout the technique as foolproof, the best possible way to match crime scene body-fluid samples to a particular victim or suspect.

DNA: THE BUILDING BLOCKS OF LIFE

DNA—deoxyribonucleic acid—often is called the genetic blueprint of all life. It is a double-stranded chain of molecules—often called a double helix—that winds its way through the nucleus of every nucleated cell in every organism (except a few viruses, which do not have DNA). From organism to organism, the sequence in which molecules form the DNA chain differs. Within species, the number of similarities is far greater than the number of differences, but because of the uniqueness of DNA, theoretically, no two people are exactly the same, save for identical twins.

The two DNA strands are made up of only four chemical building blocks—A, which stands for adenine; C, for cytosine; G, for guanine; and T, for thymine. They are strung together in a never-ending sequence. The chemicals on the two strands always go together; A always joins with T, C always joins with G. So, if one strand were AAGCTGA, the other would be TTCGACT. Figure 5.1 shows the DNA double helix.

Huge pieces of DNA are the same for every human being,

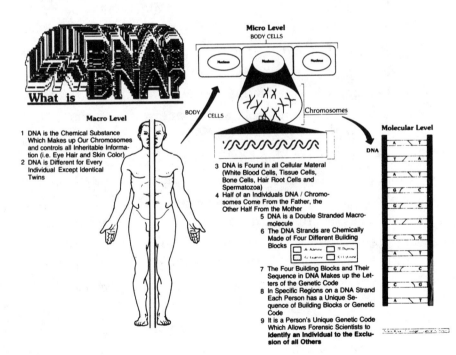

FIGURE 5.1
DNA: The Genetic Blueprint of Life

because each of us has the same body components: hands, feet, heart, brain, sinuses, etc. A few sections of human DNA vary from individual to individual, however. These *polymorphic*— variable in shape—DNA segments are what makes each of us unique. To basic scientists, studying DNA throughout the 1960s and 1970s to understand the differences in species, these polymorphic strands became known as *junk DNA.* To a few scientists, however, such as Ray White of the University of Utah and Alec Jeffreys of the University of Leicester in England, these polymorphic segments looked like a promising way to identify individuals. White and Jeffreys first published their thoughts in the early 1980s.

Both the White and the Jeffreys methods for using polymorphic strands as an identifying marker—now the Lifecodes DNA-Print and the Cellmark DNA Fingerprint, respectively— individualize by using length polymorphism.

This technique is called *restriction fragment length polymorphism* (*RFLP*). To do this, a relatively large segment of pure DNA must be extracted from each biological specimen. The DNA in each person is the same in every cell, and the polymorphic piece of the chain can be obtained from any tissue sample and from many body fluids. Finding the polymorphic chain is easy when the specimen is blood deliberately drawn for the purpose of testing, but it is much harder when the specimen is a bloodstain or a semen stain, semen from a vaginal swab, fresh tissue, bone marrow or hard bone, an autopsy specimen, or even saliva (and possibly urine). It often is impossible for a lab to give a definitive answer based on such specimens, because the specimens either are not large enough or were degraded, usually by bacteria, from having been left out in the elements too long.

If DNA can be extracted from a specimen, it is then mixed with a *restriction enzyme* that cuts the DNA chain at specific sites. The restriction fragments that are created by this process vary in length, and a few of them contain the polymorphic DNA segment.

Next, the fragments are sorted by length, using a technique called *gel electrophoresis:* The fragments are put on a gel and an electrical current is applied, causing the fragments to move toward the positive electrode. Shorter fragments move across the gel more quickly than do longer fragments and, after a short time, the fragments have lined up on the gel according to size. After the fragments have paraded across the gel, the formation is transferred to a nylon membrane called a *blot.*

At this point, a genetic probe—a particular type of DNA molecule—is applied to the membrane. This probe seeks out the fragments that have the polymorphic DNA segments and attaches onto them. Because the probes are radioactive, when an X-ray photograph—known as an *autoradiograph*—of the membrane is taken, the pieces of DNA that the probes have attached to will appear as dark patches. The X-ray photograph is known as a *DNA print.* This process is shown in Figure 5.2.

By comparing the prints from different specimens to each other, scientists can tell whether the specimens match. In a paternity test, specimens of blood are drawn from the mother, the offspring, and the putative father. Theoretically, the offspring should have a print that half matches the mother's and half matches the father's. In 1965, a doctor in Georgia apparently sold a female newborn to an Ohio couple. After the infant, Jane Blasio, became an adult and knew about her adoption, she decided to use DNA tests in an effort to determine her birth mother. Blasio discovered that her possible birth mother, Kitty Self, had died in 1987. In 1997, Jane Blasio had a DNA test performed on Jamie Goss, a young man she believed to be her half brother. Unfortunately for Blasio, the tests were inconclusive. However, DNA testing has linked two other women (sold by the same doctor in the 1960s) to their birth mothers, who still live in the area around McDaysvill, Georgia.

In a criminal investigation—usually a rape or murder—the specimens usually are blood from the victim, blood from the

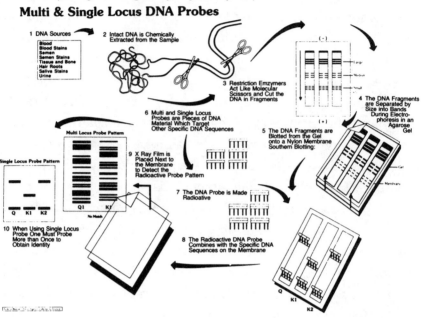

FIGURE 5.2
DNA Probes Used in Criminal Investigations

accused, and the best two or three specimens from the crime scene, such as a bloodstained item of clothing the accused was wearing, or a semen stain or swab in the case of a rape. If the DNA print of the sample matches the print of the accused's blood, scientists from the DNA-typing laboratories are willing to testify that the chances of the DNA on the sample coming from the accused are very great. These scientists argue that the chances are usually one in millions or billions that two individuals would present an identical DNA print; the odds vary, depending on the population in question because there are racial and gender differences in polymorphic strands, and various features are more common or less so. Thus, these scientists believe there is virtual certainty that the accused is guilty. This population-statistic theory has been called into question, however, as discussed in a later section.

In addition to the restriction-fragment length polymorphism technique developed by White and by Jeffreys, there is a second analytical technique, called allele-specific testing. Forensic Science Associates of California uses a test developed by the Cetus Corporation, which determines whether certain polymorphic DNA segments—called alleles—are present in a sample. This produces a much less specific identification of the DNA because large numbers of people may have any one particular allele present in their DNA. By probing for a large number of alleles, however, it is possible to say that a DNA sample came from one particular individual out of 10,000 or 100,000.

Although saying that a particular DNA print may be replicated in 1 in 100,000 individuals within a certain population is by no means saying that the print is unique, when combined with other evidence, it can be a powerful piece of a prosecution's case, not unlike the carpet in the Wayne Williams case.

The technique is especially useful when looking at small DNA samples. Because DNA deteriorates so rapidly, it is not always possible to get a large enough sample to do meaningful testing using the Cellmark or Lifecodes techniques.

In the Cetus test, DNA is purified, then "amplified," using a process called polymerase chain reaction (PCR). By heating and cooling the DNA with an enzyme called DNA polymerase, scientists can extract a few copies of a particular allele and increase it to as many as 10 million copies, sufficient to allow allele-specific probes to do their work. The amplified DNA is "spotted" onto a membrane, and allele-specific probes are added. These probes bind to different spots when that particular allele is present.

In 1988, researchers from Cetus and the University of California at Berkeley showed that there is enough DNA in a single hair to be amplified and subsequently analyzed. Although DNA from inside a hair shaft is from the mitochondria—the area of a cell that transforms the food we eat into a form of energy that the cell can use—and cannot be typed sufficiently to identify a person, the DNA from hair-follicle cells attached to hair that has been pulled out can be typed.

This use of hair follicles for PCR-based allele-specific analysis is the technique most often used to make DNA-based determinations of skeletal remains. Hence, it is the technique used on the remains in mass graves around the world, where ethnic cleansing, military abduction and execution, and other kinds of atrocities have increasingly been investigated by forensic teams in the 1990s.

OF PROBES AND STATISTICS

The idea of genetic probes is not new. Many scientists spent the decade of the 1980s searching for probes that can be sent into samples of DNA to seek out specific sites where genetic information is sent out that causes hereditary diseases, such as sickle-cell anemia, Tay-Sachs disease, or Huntington's chorea. The hope is that after these genetic disease markers have been discovered, scientists and physicians will be able to replace defective genetic material with material that does not carry the information making an individual prone to the disease, by using genetic-engineering techniques.

In 1980, Ray White discovered that variations in human DNA can be detected by the technique of restriction fragment length polymorphism (RFLP) analysis, and he described the first identification probe to do it.

Piggybacking on this discovery, Alec Jeffreys, working in Leicester, England, discovered that every human gene contains a number of "minisatellite" regions, which repeat and contain core sequences. It is the number of these minisatellite regions and, thus, the length of the fragment, that differentiates each individual. In 1983, Jeffreys produced the first probe that would bind to an entire core sequence in a minisatellite region. The X-ray picture produced by this *multilocus probe* is an image of 30 to 40 dark bands, similar to a supermarket bar code. The picture of this multilocus probe is the DNA Fingerprint that Jeffreys licensed to Cellmark, a British company, and its U.S. subsidiary in Maryland. Since 1983, Jeffreys has produced a number of probes that bind to single sites within the minisatellite regions and yield banding patterns that are easier to read. Cellmark currently uses these probes in its forensic work.

At Lifecodes, a different set of probes is used to produce the DNA-Print pictures. These probes recognize variable number tandem repeat (VNTR) sequences that appear in the DNA. VNTRs are similar to minisatellites, but while there are many minisatellites of the same type in human DNA, there is only one VNTR of each type. Lifecodes probes four specific VNTR sites in each sample. The company has a large database of VNTR information that it has used to compute the probability of each possible pattern for each VNTR appearing in various ethnic and racial groups.

The Cellmark multilocus probe or single-locus probes show up as a single print that its analysts say is unique at a level of one person in some 10 billion (there are about 5 billion people on earth). Each Lifecodes VNTR probe has a probability of one person in hundreds having the same pattern, but when the four patterns from the four different probes are multiplied, Lifecodes analysts say that their test is unique to a level of one person in billions.

The FBI, which began doing forensic DNA typing in its laboratory in 1990, uses a different set of markers. Until 1997, commercial laboratories, the FBI, and any state or local forensic labs that did DNA typing all used four sites on the DNA to try to match. In 1997, the FBI added two sites, and it was hoping to add two more sites by the end of 1998. With an eight-site match, and even with a six-site match, the FBI now believes that it can have its forensic experts say with certainty in some cases that there is no other person in the United States other than the defendant (and possibly the defendant's identical twin) who could have left the blood or semen or hair follicles associated with a crime.

A HARD LESSON

When the Lifecodes scientists came back to Tim Berry, the Orange County, Florida, assistant district attorney who was preparing to prosecute Andrews for the first of the two rapes in which he had been charged, Berry thought he had it made. He would get a definite conviction and make legal history.

In the early morning of May 9, 1986, Nancy Hodge, a 27-year-old computer operator at Disney World, had been taking out her contact lenses before going to bed when she heard a noise in the hallway outside her bathroom. She turned and—in the few seconds before she was thrown to the floor, beaten, and raped—saw the face of her assailant.

Over the next ten months, other women in the southeast part of Orlando were raped. In addition, police responded to a rash of calls about prowlers, men breaking into women's homes, and attempted assaults and rapes. Police believed that the same man was responsible for all the incidents.

On February 22, 1987, a 27-year-old mother of two was awakened from her sleep and raped while her children slept in the room next door. Police already were staking out the neighborhoods where they felt the next incident might occur and, barely a week after the assault on the young mother, a man was arrested by police responding to a report of a prowler.

The man was Tommie Lee Andrews. His fingerprints matched two latent prints lifted from the window sash of the young mother's house, and Nancy Hodge identified his photograph immediately as the man who had raped her nearly a year before.

To prepare the jury in the Hodge case for the presentation of the DNA evidence, Berry first called David Housman, a molecular biologist at MIT, who explained to the four-man, two-woman jury the theoretical basics of how DNA works. Then Nancy Hodge took the stand and identified Andrews as her assailant.

Berry then put on his key genetic witnesses, Michael Baird and Alan Giusti of Lifecodes, who had done the genetic testing. After walking the jury through the DNA-typing techniques, Baird and Giusti displayed on a light box how the autoradiograph of the DNA from Andrew's blood clearly matched that of the vaginal swab taken from Hodge 17 months earlier.

Then Berry's case hit a snag. He asked Baird to explain how Lifecodes derived its statistical assertion that Andrew's DNA was unique to a level of 1 in 10 billion. The defense objected, arguing that at no time had the prosecution brought up the subject of statistical probabilities. In essence, defense attorneys said, Berry had put the cart before the horse. Without telling the jury that DNA profiling depended on statistical probabilities, his experts would be allowed only to make an assertion that their technique proved that the chance of the DNA in the crime-scene sample not being from Andrews was 1 in 10 billion. They would not be allowed to support that assertion with their statistical theory. Berry had no legal argument to back up their assertion and had to withdraw the question.

The novelty of the technology had tripped Berry up.

Many people have at least a passing familiarity with the concept of blood type—A, B, AB, or O. Most of us know our own blood type, from having donated blood, having been typed before surgery, having been in the military, or other reasons. So when a forensic serologist—the crime-lab technician who deals with blood samples—testifies that the defendant has a particular blood type and that this blood type is found in a certain percentage of the population, jurors at least can understand the numbers, even though they may not understand the particulars of how those population statistics are derived.

DNA technology is unfamiliar, complex, and novel to jurors,

however, and the population statistics on which DNA-typing scientists base their findings of probability also are far more complex. The DNA-Print technology used by Lifecodes, the DNA Fingerprint technology used by Cellmark, and the other particular techniques used by the FBI and other forensic labs that do DNA typing are all composites—a number of probes with various statistical probabilities—that then must be multiplied to produce a final statistical probability (not unlike the case of the rug fibers in Wayne Williams's trial). The chance of any one particular probe showing a print at a certain location on the autoradiograph usually is one in thousands and varies among different subpopulations, based on race, ethnicity, and gender.

What Berry should have done was to have the Lifecodes scientists describe the population-statistics theory at the same time they discussed the DNA-probe technology and techniques. Then their assertion would have carried more weight. Without the proper theoretical foundation, the assertion was nearly incomprehensible to the lay jurors. In addition, the defense's ability to have the statistics deemed inadmissible must have made the scientific argument seem more spurious to jurors.

Andrews denied that he had left his apartment on the night Hodge was raped, and the case turned into a rather routine rape case: the word of the victim against the word of the accused. The jury came back deadlocked at 11 to 1, with an engineer refusing to accept the science involved. The judge declared a mistrial.

Two weeks later, however, Andrews stood trial again, this time for the February 1987 rape. In this trial, the prosecution was able to establish a precedent for making statistical analysis of forensic results. By showing that statistical analysis always is used to determine the probability that the evidence is what an expert asserts it to be, the prosecution could get a judge to allow a new scientific technique to be used and analyzed statistically. It didn't take long for the jury to convict Andrews. He was sentenced to 22 years in prison. A couple of months later, in

February 1988, Andrews was retried for the Hodge rape and convicted. He received three concurrent terms for sexual battery, armed burglary, and aggravated battery, the longest being 78 years. These terms are to be served after he finishes his other 22-year term.

A Natural Progression

DNA typing can be seen as the latest step in the quest of forensic scientists to individualize blood or body-fluid samples.

The journey began at the turn of the century, when Paul Uhlenhuth, an assistant at the Hygienic Institute of the University of Greifswald, Germany, wrote a scientific paper entitled, "A Method for Investigation of Different Types of Blood, Especially for the Differential Diagnosis of Human Blood," in which he wrote, "I have succeeded, taking blood of men, horses and cattle dissolved in physiological NaCL [salt] and dried on a board for four weeks, in identifying the human blood at once by means of my serum."

Until that time, it was impossible to tell whether a dried stain was blood, or whether the obvious bloodstain on a person's clothing or hands came from a victim of assault, murder, or rape or merely from a chicken the person had killed for supper.

Since 1890, when Emil von Behring discovered the principle of antitoxins, scientists throughout Europe had searched for ways to protect people and livestock from various diseases. Von Behring discovered that when a cow was inoculated with a small amount of diphtheria toxin, the blood serum—the watery component of blood—formed defensive antitoxins. Further studies showed that blood formed defenses against all kinds of foreign proteins, including other animals' milk and other animals' blood. When there were foreign proteins present, the serum clouded over; when it was mixed with blood from its own species, it remained clear. The serum worked not only with fresh blood, but also with old, dried bloodstains.

Knowing that a bloodstain is human blood gets the police and forensic experts only a short way down the road, however; the greater distance must be traveled by individualizing the stain. Over time, scientists have determined a number of genetic markers to do this; these markers fall into five major categories:

1. Blood groups

2. Isoenzymes

3. Serum groups

4. Hemoglobin variants

5. The HLA system

Blood Groups

There are about 15 well-established blood-group systems, but only a few are used frequently for analyzing bloodstains in forensic cases. Blood groups are defined by antigens on the surface of the red cells. An *antigen* is a substance that reacts to a particular protein.

The ABO blood-group system probably is the most well known; this was the first system described, just after the turn of the century. Whole-blood typing is done by separating the serum (also known as *plasma*) from the whole blood and testing the serum with anti-A, anti-B, and anti-H serums. The reactions determine whether the blood type is A, B, AB, or O. ABO testing of bloodstains also can be done with other methods, some of which have been used since the 1930s.

It also was determined in the 1930s that many body-fluid stains contain soluble ABO blood-group substances; over time, it has been determined that the majority of people do secrete ABO substances into their semen, saliva, and many other bodily fluids. These people are known as *secretors*. The FBI laboratory's serology unit, which examines some 25,000 samples each year in 2,300 cases, uses three blood group tests: ABO, Rhesus (Rh), and Lewis (Le).

Isoenzymes

Isoenzyme systems—multiple molecular forms of a particular enzyme—are coded for at particular genetic loci in some individuals. In other words, some people have within their genetic makeup at particular points on every gene particular molecular forms of a number of different enzymes. Everyone has some of these enzymes, though no one has all of them. The variation of these enzymes can be compared from one blood sample to another, or from a blood sample to body-fluid samples or blood- or body-fluid stains.

The FBI routinely tests for nine isoenzymes: phosphogluco-mutase (PGM), esterase D (EsD), glyoxylase I (GLO 1), erthrocyte acid phosphatase (EAP), adenosine deaminase (ADA), adenylate kinase (AK), carbonic anhydrase (CAII), peptidase A (Pep A), and glucose-6-phosphate dehydrogenase (G6PD).

Serum Groups, Hemoglobin Variants, and the HLA System

It is also possible to compare blood or body-fluid samples or stains by hemoglobin variants, other serum-group systems such as haptoglobin and transferrin, and the HLA system. The FBI routinely tests for haptoglobin (Hp), transferrin (Tf), and group-specific component (Gc). Performing each of these genetic-marker tests on samples from a crime scene and on samples from suspected individuals continually narrows the field of possible matches; that is, when particular types are not possessed by an individual, this lack immediately excludes him or her as a possible source of the crime-scene sample.

If a suspected individual possesses all the types tested for and found in a crime-scene sample, the possibility of that person being the actual depositor of the sample can be computed because population-distribution data about these systems have been kept for many years and broken into subpopulations by racial and gender factors.

The investigation of sexual assault has been helped by some laboratory techniques that depend not on blood, but on the specifics of semen.

One such technique is an *assay,* or test, that searches for anti-sperm antibodies, which often can be found in men who have had a vasectomy; they develop antibodies to their own sperm. The test was first used in 1983 to convict a San Bernardino County, California, police officer of raping and murdering a young woman he had stopped for speeding.

Another test uses the monoclonal antibody MHS-5, a substance that detects the presence of seminal fluid in unspecific samples by searching for the specific protein produced only in the seminal vesicles, the part of the male testes that produces semen.

DETERMINING THE SCIENTIFIC ACCEPTABILITY OF DNA TYPING

Before Berry could introduce his DNA-typing evidence in the Andrews cases, the judge conducted what has come to be known as a Frye hearing, also called a "general-acceptance test" hearing, named after the 1923 U.S. Court of Appeals case in which a judge first ruled that a scientific process used in criminal investigation had enough acceptance in the general scientific community to make it admissible in court. In *Frye v. United States,* the Court of Appeals for the District of Columbia ruled that

> Just when a scientific principle or discovery crosses the
> line between the experimental and demonstrable stages is
> difficult to define. Somewhere in this twilight zone the
> evidential force of the principle must be recognized, and

while courts will go a long way in admitting expert testimony deduced from a well-recognized scientific principle or discovery, the thing from which the deduction is made must be sufficiently established to have gained general acceptance in the particular field in which it belongs.

Similar nonjury hearings have been held regarding the general acceptability of DNA typing in a number of states. At the end of 1988, Lifecodes reported that it had never had its evidence ruled inadmissible. All that would change in 1989, in a case in its own state, New York.

Events on the DNA-typing front moved quickly in the late 1980s through the state of New York. In late 1988, prosecutors in Queens County, New York, secured that state's first conviction using DNA typing, and prosecutors in Albany County were pressing two cases, hoping they could be used to test the Queens standard throughout the state court system.

In addition, a state assembly committee held hearings on whether to legislate regulations of the practice. Defense attorneys argued that while DNA typing is a valid technique in theory, there is no legislated quality control over the commercial laboratories conducting the tests. They noted that the state health department had not yet set quality-assurance standards for DNA typing as a diagnostic tool, which is a technique identical to forensic DNA typing.

The 1988 Frye hearing, before Albany County Judge Joseph Harris, cost the county public defender over $50,000, including $20,000 for expert witnesses, and the proceeding transcript ran over 1,000 pages. Judge Harris hoped that his hearing would have precedent value; he ruled that the technology is "the commercialization of long-standing principles that have been adopted by the scientific community." He agreed, however, that there ought to be some sort of licensing of commercial facilities to stem the proliferation of "street-corner" laboratories.

Lifecodes had its big setback beginning in May 1989, when expert witnesses for both the prosecution and the defense in a Bronx, New York, case held an unprecedented out-of-court miniseminar and issued a two-page consensus statement that said, in part, "Overall, the DNA data in this case are not scientifically reliable enough" to reach a conclusion. The scientists continued, "If [these] data were submitted to a peer reviewed journal in support of a conclusion, it would not be accepted."

The case involved Jose Castro, a 38-year-old Bronx resident accused of murdering a neighbor, Vilma Ponce, and her 2-year-old daughter. In addition to an identification of Castro by Ponce's common-law husband, a spot of blood on Castro's watch was determined by Lifecodes to be that of Vilma Ponce. In its July 22, 1987, forensic report, Lifecodes wrote, "The DNA-Print pattern from the blood of Ponce matches that of the watch with three DNA probes. The frequency of these patterns in the general public is 1:189,200,000."

When the case came to trial in February 1989, the defense attorney asked for a Frye hearing on the validity of the DNA evidence (a Frye hearing can be held in any jurisdiction within a state if no state court of appeals has held the scientific evidence valid). The attorney turned to two other attorneys, Barry Scheck (a professor at Yeshiva University's Cardozo School of Law) and Peter Neufield, for assistance. Scheck and Neufield had recently been to a conference on forensic DNA typing at the Cold Spring Harbor Laboratories on Long Island, at which the potential problems of the technology had been discussed.

One participant at that seminar was Eric Lander, a human geneticist at the Whitehead Institute in Cambridge, Massachusetts (and, as such, a colleague of David Baltimore, whose colleague, Thereza Imanishi-Kari, had had her lab books reviewed by Secret Service forensic writing experts looking for scientific fraud in a paper published in the journal *Cell;* see Chapter 3).

At the Cold Spring Harbor conference, Lander had challenged Michael Baird of Lifecodes about a presentation Baird made, an incident that had made Scheck and Neufield sit up and take notice. As Lander told *Science* magazine for an article in the News and Comment section of its June 2, 1989, issue, "Michael Baird showed a slide of an autorad with two lanes" (an autoradiograph of the DNA pattern from two different samples in "lanes" next to each other). "The bands didn't line up, but he called them a match anyway, saying that one lane ran faster than the other" (a phenomenon known in such experiments as "band shifting"). "I got up and said, 'How can you tell it's band shifting? Where are the internal controls?'"

A control study can be run in each lane, showing a fragment of a known length. If there is a band shift, it can be determined against these control lengths.

Lander was equally unimpressed with a presentation made by a participant from Cellmark. He told *Science* that in the Cellmark presentation, "A match was called even though all bands didn't correspond, but for a different problem, DNA degradation. No controls there."

Neufield contacted Lander while preparing for the Castro case, and after Lander reviewed all the material presented and transcripts of the Frye hearing, he agreed to testify; he spent six days on the witness stand and prepared a 50-page report.

In early May, 1989, Lander and the seven other expert witnesses who participated in the Frye hearing for the Castro case received permission from the judge and lawyers to hold the miniseminar without any of the legal personnel attending, a truly unique occurrence. In their consensus statement, the experts did not put the blame on Lifecodes, but rather on the adversarial nature of the legal system and its inability to deal effectively with complex science, as well as on themselves as leaders in the scientific community for not pushing harder for universal standards in forensic use of the technique. They wrote, "All experts have agreed that the Frye test and the setting of the adversary system may not be the most appropriate method for reaching scientific consensus" and that "there is a need to reach general scientific agreement about appropriate standards for the practice of forensic DNA typing."

The experts called on the National Academy of Sciences to study the issue. The Congressional Office of Technology Assessment (OTA) initiated a study of the forensic use of DNA typing early in 1989 and issued its report in the winter of 1990. Lander was a key expert helping to write that report.

On August 14, 1989, Acting Justice Gerald Sheindlin of the New York State Supreme Court ruled that DNA typing, in theory, is an admissible scientific technique. Sheindlin ruled, however, that the Lifecodes tests were admissible as *exclusionary tests*—to show that the blood on the watch was not Castro's—but inadmissible as *inclusionary tests*—to show that the blood on the watch was Ponce's. He wrote, "The testing laboratory failed in several major respects to use the generally accepted scientific techniques and experiments for obtaining reliable results, within a reasonable degree of scientific certainty." Judge Sheindlin suggested that attorneys who had defended people convicted on the basis of DNA-typing evidence examine the trial transcripts to see whether they could have the convictions overturned on appeal by using his ruling as precedent.

In addition to criticizing sloppy laboratory techniques, Lander also has questioned the assumptions about population genetics that led the expert witnesses from the commercial labs to testify that the chance of a person other than the defendant matching the DNA-typing test is one in hundreds of millions or even one in billions.

In their article on individualization of blood and body fluids for the American Chemical Society's book *Forensics Science*, Robert Gaensslen, then professor of forensic science at the University of New Haven; Peter Desio, also of the University of New

Haven; and Henry Lee, then chief of the Connecticut State Police Forensics Laboratory, warn about the use of population statistics:

> Computation of the approximate occurrence of a type or set of types in a population is possible because population-distribution data have been collected for many systems in many different populations, including U.S. populations. . . . There are choices to be made in the presentation of this kind of information to a court. . . . There are different ways of looking at population frequency data, and care must be taken not to make this information appear to have more significance than it does.

Judge Sheindlin's ruling in the Castro case shows a level of distinction between generally accepted scientific practice and common forensic-science practice not always adhered to in Frye hearings. Too often, argues Bert Black, an attorney in Baltimore, judges accept the expertise of forensic-lab technicians and their arguments about the ability of their techniques to meet the requirements of general scientific acceptance.

A case in point, he argued in a scholarly article for *Science* magazine, is the issue of the so-called multisystem technique of blood typing by electrophoresis. This technique was designed by police forensic-lab scientists to identify several genetic markers from a single crime-scene blood sample by separating the proteins using electrophoresis and sequentially staining the sample for different tests. In addition to questions about the reliability of sequential tests, there are also questions about contamination of crime-scene samples and degradation of the same over time.

In 1986, the Michigan Supreme Court, in *People v. Young,* overturned a rape–murder conviction that had depended on the multisystem test. First, unlike the Kansas Supreme Court—which had ruled the evidence from the multisystem test admissible in *State v. Washington,* a 1981 rape–murder case it upheld— the Michigan court did not accept the testimony of a crime lab technician, but demanded expert testimony from research scientists other than those who had developed the process.

"The scientific tradition expects independent verification of new procedures," the court wrote, noting that in the case of the multisystem test, the only data supporting it were in an unpublished report written by the test's developer and sponsored in the 1970s by the Law Enforcement Assistance Administration (LEAA), an arm of the Justice Department disbanded in the 1980s.

Despite the challenge in the Castro case, in the spring of 1989, the FBI began accepting requests from law-enforcement agencies across the country to conduct DNA-typing tests. Because of the anticipated crush of cases, the FBI limited samples to those taken from violent crimes where there is a specific suspect, or from cases that could be one in a series of rapes, murders, or child molestations, even if there is no current suspect. The agency also began training 60 technicians from local or state forensic laboratories and hopes to increase the number of local technicians trained each year.

Despite the FBI putting its imprimatur on the technique of forensic DNA typing, the process has continued to generate controversy throughout the 1990s. In December 1991, the debate exploded in the pages of *Science* magazine, the journal of the American Association for the Advancement of Science. In that magazine, Richard Lewontin of Harvard University and Daniel Hartl of Washington University, two eminent population geneticists, argued that the kinds of statements made by laboratory personnel regarding the odds of a particular crime-scene sample coming from a particular individual were "unjustifiable," because there were no reliable data on genetic variation among ethnic groups that could lead to such statements. Ranajit Chakraborty of the University of Texas, Kenneth Kidd of Yale, and Thomas Caskey of Baylor College of Medicine argued in the same issue that it really doesn't matter whether the odds are 1 in 500,000 or 1 in 5,000,000.

In April 1992, the National Academy of Sciences (NAS) issued the report of its panel that began looking into the question in 1989. That panel said forensic DNA typing could be useful, but only if labs performing the work adhered to the highest scientific standards, and if there was proficiency testing. The report did not say who should (a) conduct those proficiency tests, (b) determine exactly what the standards should be, or (c) rule on which labs met those standards.

By 1992, it looked as though a full-scale backlash against forensic DNA typing was in force. Whereas prosecutors had used the evidence to overwhelm a number of defendants in the late 1980s and early 1990s—most of whom were indigent and represented by public defenders, with few or no resources to hire experts to counter prosecution witnesses—during the mid- and late 1990s, scientists were increasingly calling the procedure and especially the population geneticists into question.

After the 1989 Castro case had teamed up Peter Neufield and Barry Scheck, the two began using forensic DNA typing in an effort to free those they believed had been convicted unjustly. In 1992, they started the Innocence Project, and by 1997, the project counted nearly 50 victories, often overturning convictions where individuals had spent more than 15 years in prison, using DNA typing of old crime-scene samples.

Neufield and Scheck also teamed up to put the greatest assault on forensic DNA typing in 1995, when they joined the defense team of O. J. Simpson. Scheck conduced the cross-examinations of most of the prosecution's forensic witnesses, and he enlisted Henry Lee, then Chief of the Connecticut State Police Crime Lab, as Simpson's senior expert in an attempt to try to beat back the prosecution's forensic case. Anticipating that Simpson's high-priced and high-powered legal team would put a full-court press on the forensic evidence after Simpson's arrest in 1994 for murdering his ex-wife and her friend, many former combatants in the DNA wars came to a truce and basically agreed that the technique was sound if a laboratory could prove that it was proficient at conducting the tests. In October 1994, a paper in the British journal *Nature* making that point was co-authored by Eric Lander of MIT, one of forensic DNA typing's greatest critics, and Bruce Budowie, a chief scientists for the FBI. The two men wrote in that paper that "the DNA fingerprint wars are over."

By throwing doubt on every aspect of the Simpson prosecution's forensic case, from the collection of samples to the chain of custody to the handling of physical evidence—the famous rumpled pants that may have transferred a stain from one place to another—to the population genetics, Scheck and Lee were able to overwhelm the prosecution experts. Simpson was found not guilty; what role the forensic evidence played is hard to determine, given the racial overtones of the case.

FREEING THE INNOCENT

While DNA as a tool of *inclusion*—trying to prove guilt—came under heavy attack throughout the 1980s, the technique became increasingly used as a tool of *exclusion*—proving that an individual could not have been guilty because blood or semen samples conclusively do not match. From its beginning in 1992, the Innocence Project focused on cases where men convicted of murder or rape felt they had not been adequately represented— usually by young, underpaid public defenders whose offices have few resources for expert witnesses—and where the prosecution evidence was flawed or inadequate to establish the individual's identity. Usually, the key piece of evidence was eyewitness identification, which experts have shown time and time again to be unreliable.

The latest victory for the Innocence Project came with the December 2, 1997, release of Dale and Ronnie Mahan, after they spent nearly 14 years in prison on charges of rape. The victim said she was raped by two men on November 30, 1983, after being abducted in a mall. The two men wore ski masks, but the victim identified the Mahan brothers from photographs, saying they lifted those masks a number of times, revealing their faces up to the bridges of their noses, in order to drink and smoke.

She also said the younger-appearing man stuttered. Dale Mahan, the younger brother, stutters. The brothers maintained that they were in a bar celebrating with family at the time of the attack. The brothers also have a fairly rare blood type, which on standard serology tests in 1984, before DNA technology became available, was shown to be "consistent with" the semen found on the victim.

Three weeks after the Mahans were freed, Arthur Green, Jr., an assistant district attorney in Jefferson, Alabama, said he planned to retry the brothers, despite three pieces of evidence that intertwine to yield an exculpatory conclusion for the brothers. First, the semen found on the victim did not come from either of them. Second, it also did not come from the woman's husband at the time, from whom she was divorced later, long before the brothers were freed. The victim's ex-husband had been killed in 1992 in a police incident, but Mississippi officials had kept blood samples from the autopsy, and Alabama investigators performed the DNA analysis on the late ex-husband's blood in December 1997.

Third, the woman has since acknowledged that on the day of the alleged attack, she not only had sex with her husband in the morning, but also with a boyfriend in the afternoon. DNA testing shows that the semen found on the woman was from her boyfriend. In addition, the woman has a history of mental instability, with a number of hospitalizations.

If the Mahan brothers are retried, they may argue that the woman had a rough sexual encounter with her boyfriend and manufactured the story of the abduction and rape to account to her husband for cuts and bruises. Green believes in his witness, however, and he plans to try the case a second time, using her eyewitness account as his strongest evidence.

The Mahan brothers' original attorney, Dan C. King, currently a judge, told *The New York Times,* "If these things were told to us at the first trial, these boys never would have been convicted." King's statement echoes those of other attorneys who have represented defendants later freed by the Innocence Project. In fact, in April of 1993, John E. Kloch, Commonwealth Attorney for Alexandria, Virginia, petitioned the court to release Walter Snyder, who had served six and one-half years on a rape charge, citing DNA typing that said Snyder's DNA did not match that of semen samples. "Had that evidence been presented at trial, he would have been acquitted," Kloch said.

JURY BELIEF IN A WITNESS OVER DNA

In a 1990 Connecticut case, prosecutors won a victory on the basis of a victim identification despite the testimony of an FBI DNA expert that the defendant's DNA did not match that of the semen found on the rape victim.

To be sure, the case of Ricky Hammond was contentious from the beginning. Hammond, of Bristol, was accused of raping the victim on November 30, 1987. DNA samples taken from semen samples on the victim's underwear prior to Hammond's 1990 trial showed DNA that did not match his.

Yet the victim picked him out of a photo lineup three times. Although she had never met him, she described in meticulous detail the conditions of his messy car, down to tears in a child's car seat and a watch that hung from his console gear shift. She described mannerisms that he displayed in open court. Hammond was convicted.

In 1991, Hammond's appeal cited the fact that Assistant State Attorney John Malone had argued to the jury about a number of scientific misadventures throughout history, including the sinking of the *Titanic* and the explosion of the space shuttle *Challenger,* but had not introduced properly during the trial the notion that DNA typing technology was not infallible. The prosecution argued at appeal, but not during trial, that enzymes from laundry detergent can contaminate DNA samples.

Finally, in late October, 1992, after a fourth round of DNA testing, Judge Richard Damiani ordered Hammond released on a promise to appear in court for a new trial. Hammond's attorneys—Todd Fernow, a lawyer with the University of Connecticut Legal Clinic, and a group of law students—handled Hammond's appeal and motions for a retrial.

The fourth set of DNA samples was both more conclusive and less conclusive than the previous sets. The samples were taken from vaginal swabs, rather than from stains on underwear; but because of degradation of the sample over time, the most sensitive DNA tests could not be performed, and the results were only good enough to show that the semen sample came from a person whose DNA could be limited to about 1 in 50 million Americans. More important, however, the results were good enough to exclude Hammond. The case later settled in a plea agreement, with Hammond given credit for the time he had served: over two years.

In all of these cases where DNA technology shows that semen samples were not left by defendants, prosecutors turn to two possibilities: (1) the defendant did not ejaculate, or (2) as is becoming more common, the rapist used a condom.

MASS TESTING AND CIVIL LIBERTIES

A much larger question than the legal and regulatory ins and outs of testing is how the country will reconcile testing with fundamental constitutional liberties. Alec Jeffreys's DNA Fingerprint test became famous in 1986 when it was used—first to clear a suspect, then to trap the killer—in the so-called Black Pad

Murders that occurred in Narborough, England, about 10 miles west of his office at the University of Leicester. In November, 1983, 15-year-old Lynda Mann had been raped and killed on her way to a babysitting job in the village of Enderby a short walk away from where she lived. On July 31, 1986, another 15-year-old, Dawn Ainsworth, was raped and strangled on her way home to Enderby. Chief Superintendent David Baker found striking similarities in the two murders: The two girls were both 15, attended school together, were raped and murdered within a mile of each other, and had similar physical appearances. He assigned Anthony Painter, a detective superintendent, to head up the Ainsworth murder investigation. Three years earlier, police had taken some 5,000 statements after Lynda Mann's death and had come up empty-handed; Painter, however, arrested a suspect in less than a week: Richard Buckland, a 17-year-old kitchen worker at a local mental hospital, confessed to killing Dawn Ainsworth.

Police asked Jeffreys to analyze samples of Buckland's blood, as well as semen stains and vaginal swabs from the two murdered girls. Although the amount of DNA in the crime-scene samples was minuscule, Jeffreys told police that while the crime-scene samples all came from the same man, they did not come from Richard Buckland. Scientists at the Home Office Forensics Laboratory confirmed the findings. Buckland was released.

Painter remained convinced, however, that DNA typing could be useful in solving the case. Based on his long years of investigating violent crimes, Painter theorized that the killer was a local man under age 30—he never has said exactly what led him to this theory—which would mean that it could be one of some 4,000 men who lived or worked in the area. Conventional serology could eliminate out 60 percent of those men, because their blood types did not match the crime-scene samples, narrowing the range to about 1,600. Painter wanted these men to have their blood samples DNA-typed by government forensic-laboratory technicians; each negative match would get the police one step closer. Because English law says that all body-fluid samples must be given voluntarily, Painter would have to get each man to volunteer his blood for the testing. He received the go-ahead from village leaders and began collecting samples on January 5, 1987.

Painter was not assuming that the killer would take the test. Rather, he was working on the assumption that the killer would deliberately duck the test—and that's exactly what happened. On September 20, 1987, police arrested Colin Pitchfork, a 27-year-old baker from Littlethorpe. Police say that Pitchfork declined a first invitation to give blood, saying that he was busy, then got a friend to donate blood, using Pitchfork's name. As so often happens, the friend boasted to drinking buddies of how he and Pitchfork had outwitted the police. Pitchfork later was convicted of the two murders, and the case became the subject of a book *The Blooding,* by Joseph Wambaugh, published in 1989.

Can such mass testing happen in the United States? Legally, in the United States, if police have probable cause, they can get a judge to issue an order compelling an individual to give a sample of his or her body fluids. Legal challenges to blood tests on Fourth Amendment grounds in the early part of this century were denied in the same way that refusal to give fingerprint evidence was denied.

Whether or not hundreds or thousands of people in the United States would volunteer for a mass blood draw is another question, and the answer probably is no, given how inured we have become to violent crime and how litigious we are in defending constitutional rights. There is precedent for large-scale testing in criminal cases, but the level of cooperation shown by the people of the Narborough area in England probably would not be shown here.

In addition, some law-enforcement officials in the United States already are beginning to argue that DNA typing, combined with powerful computer databases, could be the perfect form of universal identification.

Colorado was the first state to enact legislation forcing convicted sex offenders to submit a blood test for DNA typing in order to be released. In California, the state attorney general in early 1989 began pushing for DNA typing to be done on all people convicted of many violent crimes. Under the California plan, anyone convicted of murder, assault, rape, or other sex crimes would be required to give two blood samples and a saliva sample upon entering prison. The California attorney general, John Van de Kamp, estimated that it would cost about $3 million to set up the computer system, which would be accessible to any law-enforcement agency in the state. He also estimated that the annual cost of taking and typing some 8,250 samples per year—the expected number—would be about $2.5 million. The records would be permanent. The Van de Kamp legislation in its original form did not pass, but California did eventually set up a database of sex offenders' DNA, as did many other states.

The FBI considered asking for similar federal legislation but eventually left it up to states to enact legislation separately. Once states have done so, the FBI will collect the information in the National Crime Information Center (NCIC), an FBI-directed computer network of information about crimes throughout the country (see Chapter 8 for more detail about the NCIC). In January, 1992, the country's military services created a DNA-type database to be used in identification of soldiers killed in action or in accidents. In June 1988, using technology to recover DNA from bones, the military was able to identify the Vietnam-era "unknown" soldier as Air Force Lieutenant Michael J. Blassie, whose plane was shot down near the village of An Loc on May 11, 1972.

The U.S. Army, the lead agency in this effort, has offered assurances that the samples will be sealed and stored frozen, in order to maintain specimen integrity over a long period of time. In regard to concerns about privacy and confidentiality, the Army indicated that the records will be kept only as long as an individual is in active service or the reserves and will be destroyed upon separation from the military; further, the specimens will be used only for identification after death, not for criminal identification or paternity, unless a court orders the military to turn over the specimen.

As with other schemes for universal identification, including Social Security numbers, civil libertarians worry about the misuse of DNA typing by government officials to intimidate and spy on citizens' private lives, and they are marshaling their arguments against such use. Attorneys for the Project on Privacy and Technology of the American Civil Liberties Union (ACLU), harsh critics of the entire NCIC program, make the following argument: Whereas DNA typing as it is used now matches a specific sample to a specific suspect, the computer database will allow police to round up suspects based on cursory examination of DNA-typing tests from crime scenes.

Although a national system is still not in force, in December 1997, the first use of interstate database transfer of DNA-typing information took place, when an Indiana rape investigation, searching through the states' DNA-typing databases using the FBI link, discovered that the crime-scene samples they had in hand matched the DNA of a man serving time for rape in Illinois.

WORKING BACKWARD: NO-BODY MURDERS AND REUNITING FAMILIES

In July 1988, Lisa Tu, a 42-year-old Potomac, Maryland, resident, disappeared. Her longtime companion, 58-year-old Gregory Tu (no relation), told police that Lisa Tu had gone to visit a sick friend in San Francisco. Lisa Tu's children had spoken to the friend, however, who said that she was not sick and had not seen Lisa Tu in months. Police arrested Gregory Tu and charged him with his common-law wife's murder, saying that he killed her in their home and disposed of the body. When searching the Tu house, police found splattered blood in the finished portion of the basement. They asked Cellmark to determine whether the spattered blood was that of Lisa Tu.

The Tu case is the third so-called no-body murder in which DNA typing has been used. The first was in Oklahoma in 1987; the case ended in acquittal, despite the jury's belief that the DNA evidence was convincing, because the test itself cannot show whether the person from whom a sample comes is alive or dead, only that a sample taken from the scene of what appears to be a violent crime probably came from the missing person. The second case, in Kansas in 1988, ended in a conviction. Gregory Tu's trial was scheduled for April 1989.

Without a body, DNA investigators must work backward, in effect doing a reverse paternity (or maternity) test, to determine the DNA characteristics of the apparent victim, then match those characteristics to crime-scene samples. In the Tu case, police obtained a sample of blood from Lisa Tu's teenage son from a previous marriage, as well as from her ex-husband, Wing Lau, who currently lives in Hong Kong.

Since children obtain half of their genetic information from each parent, half the bands in the probed regions of Lisa Tu's son's blood should have matched those of his father's blood. The other half of the bands were then used as a comparison against the crime-scene samples. Because half the bands of the crime-scene samples matched those bands of Lisa Tu's son's blood that did not match his father's, genetic experts were willing to testify that the bloodstains in the basement were indeed stains of Lisa Tu's blood. Although this does not prove her dead, juries have been known to convict in no-body murders on the basis of bloodstains. Gregory Tu's lawyers challenged the admissibility of this reverse testing, although DNA typing already had been ruled admissible in Maryland courts. He was convicted.

Another example of the use of molecular evidence has occurred under the auspices of the American Association for the Advancement of Science's Committee on Scientific Freedom and

Responsibility. Mary-Clair King, professor of epidemiology at the University of California at Berkeley, described the work at the 1989 AAAS meeting: From 1976 through 1983, Argentineans lived under a brutal military rule, under which the Argentine military conducted what has come to be known as the Dirty War. More than 9,000 Argentineans were officially listed as disappeared, often taken from their homes in the middle of the night by roving bands of security forces—men driving unmarked and unlicensed cars. While many adults were tortured and murdered, and thousands of others were presumed to have been murdered, children were often "adopted" by military or government officials.

During the Dirty War, more than 200 such infant kidnappings occurred. Now, using mitochondrial DNA—which is passed on identically only from mother to child—King and her colleagues, who have been working in Argentina since shortly after the country's return to democracy in 1984, have identified 60 children, using blood samples from maternal relatives and the "adoptive" parents to prove that the children have living biological relatives. Of the cases resolved, King said that 4 children have died, 46 have been living with people involved with the military—many of whom since have been convicted of crimes relating to the reign of terror—and 8 children were living with adoptive parents who thought the children had been abandoned.

DNA Technology into the Twenty-first Century

Over the decade of its use, the pendulum of DNA-typing technology has swung from being a "miracle tool" of police science, an infallible method of putting violent criminals away, to a tool used with equal skill by forensic experts for the defense, and as a humanitarian tool by forensic scientists investigating war crimes and other crimes against humanity around the world.

As we enter the twenty-first century, the technology is

becoming increasingly sensitive. In 1997, the FBI added two more sites to the probes it uses in its Crime Lab DNA-typing technology. Measuring a known sample against an unknown sample using six sites instead of four increases the specificity with which a match can be declared to 1 in 250 million, according to FBI experts. This makes it possible, they argue, for an FBI DNA examiner to testify that, for all intents and purposes, the sample is specific to one individual in the United States (at the end of 1997 there were approximately 267 million Americans). It is hoped that by the end of 1998, and certainly by the end of 1999, the FBI will use an eight-site probe, making the match even more authoritative.

As more and more states create DNA databases, comprising known samples from violent offenders who are imprisoned or who are being released from prison, and as those states put their databases on the FBI Violent Criminal Apprehension Program (VICAP) server, more "hits" like the one between Indiana and Illinois will occur. There may come a time when, truly, a serially violent individual who rapes, assaults, and kills will be able to run but not to hide.

At the same time, courts are very cognizant of the unreliability of eyewitness testimony, even from a victim. In cases where DNA-typing evidence shows that a crime-scene sample does not match the DNA from an individual convicted of the crime, they are throwing prison doors open. Some prosecutors retry these cases; others do not. Still others settle in a plea arrangement.

On the human-rights scene, three-way DNA typing of spouses, parents, and children are allowing investigators to positively identify the remains of hundreds of victims of political genocide around the globe.

The technology is clearly here, it is getting more sophisticated, and those who use the technology for forensic uses are, increasingly, being held to the highest possible standards of laboratory science. They need to be, because people's lives and freedoms rest on their work.

6

THE MIND AND BRAIN OF VIOLENT CRIMINALS

The first body was discovered on May 13, 1984, in a remote area of southern Hillsborough County, Florida, near Tampa. It was that of a 20-year-old Laotian woman who worked as an exotic dancer at a Tampa bar. She was nude, lying face down, with her hands tied behind her. She had been strangled and had been deliberately "displayed" with her feet spread about 5 feet apart. Around her neck was a rope with a leashlike extension. She was found near an interstate highway, but she had not been killed there.

Two weeks later, in an isolated area of eastern Hillsborough County, another body was found. This was a 22-year-old California native, a Caucasian. A former prostitute, she also was found nude, but on her back, with her hands tied at her waist. Her throat had been cut, and she had been hit on the head many times with a blunt instrument. She, too, had a rope and leash around her neck and was found near an interstate. She also had been killed somewhere else and then dumped. This woman's clothing was found near her, and serology tests were performed on semen stains on her clothing.

On June 24, a third body was found, this time in the southeastern part of the county. Another Caucasian woman, she turned out to be a 22-year-old Tampa native who worked on an assembly line. She was known to frequent an area in the northern part of the county that was a place for prostitutes to work looking for customers, but she had no criminal record. She was found fully clothed, and there was no ligature (rope, leash, or cord) around her neck. She was found in an orange grove, not near a highway. She probably had been dead for about two weeks, as opposed to the other two women, who were found within one to three days after their deaths.

Three and a half months later, on October 7, a fourth body was discovered, this time a black 18-year-old woman, and this time on the border of Pasco and Hillsborough Counties. She had an arrest record for prostitution and had last been seen a week earlier. Her body was found near the entrance to a cattle ranch. She had been killed by a gunshot to the neck. Her clothing was found near her body.

At this point, the Hillsborough County Sheriff's Office, which was conducting the investigation, asked the FBI to do a criminal personality profile on the first, second, and fourth murders (it was not until months later, when laboratory evidence of hairs and fibers was compared, that the third murder was recognized as part of the series). The profilers saw only enough similarity to link just the first and second cases. They furnished the sheriff's office with the following profile.

Police were probably looking for a white man in his mid-20s with a macho image. He probably was divorced and had difficulty holding a job. He drove a flashy car and was likely to carry weapons. He would be generally assaultive, with a tendency to both mentally and physically taunt and torture people. He chose his victims randomly and chose victims who were susceptible to easy approach. He would confine his activities to a relatively bounded geographic area.

By the end of November, four other murders had been linked to the series. Police from three jurisdictions (Pasco and Hillsborough Counties and Tampa) were each investigating the murders that had occurred in their own areas, but all were in constant communication. In addition, the Tampa police were investigating the abduction and rape of a 17-year-old girl.

In December, police arrested Robert Joe Long and charged him with one of the murders. Long was eventually tied to 10 murders in the Tampa Bay area. Long was 31, divorced, and unemployed, having been fired from his last job. He was on probation for assault. He drove a red Dodge Magnum, carried a gun and a knife, lifted weights, and had tied a leash around the necks of some of his victims.

How did the FBI profiler come up with such a specific picture of the man police should be looking for? While the FBI will not discuss particular cases, members of the Behavioral Sciences Unit (BSU) in Quantico, Virginia, will talk about technique in general. In addition, since his retirement as the head of the FBI Behavioral Services Unit, John Douglas has written three successful books about his work as a "mindhunter."

While the criminalist and forensic scientist examine a crime scene looking for the most minute bits of physical evidence to identify the victim, the perpetrator, and the exact circumstances of the crime, another group of investigators—the behaviorists—look at the crime scene for evidence of motivation, such intangible factors as rage, hatred, love, or fear.

The work of behavioral investigators most often is done in

crimes of extreme violence—murder, rape, child molestation, and arson. These investigators include a handful of special agents in the FBI's BSU and dozens of police officers around the country who have been trained by the BSU agents in an 11-month program at the FBI academy in Quantico. The behaviorists seek to get into the mind of the criminal and explore his (or her) thought processes in an attempt to tell investigators what kind of person would commit such a crime and narrow the field of potential suspects by describing behavioral—and often physical, family, and lifestyle—characteristics of the person investigators should be looking for. At times, they also suggest to investigators ways to question potential suspects in order to use the suspect's emotional vulnerabilities.

The tool used by behavioral investigators to do this is known as the psychological or behavioral profile, or what the FBI now calls "criminal investigative analysis." In the past, this work was often done by psychiatrists or psychologists working as consultants to police departments. Some consulting psychiatrists and psychologists still work on profiles, although since the 1980s, more police departments that had previously used this tool now turn to the FBI or an FBI-trained police investigator in their geographic area.

The FBI's profiling generally has received high marks, although police officers are less thrilled with the results they have received from psychiatrists or psychologists. Some investigators swear by the technique's usefulness in furthering investigations, especially in gruesome, seemingly senseless murders and in serial murder and rape cases. They say that the profile often resembles someone they have already questioned but dismissed as a suspect because there just was no physical evidence—and no witnesses—to go on. Other investigators say that behavioral profiles are almost always useless; they either are too general, being based on common sense and statistical probabilities, or are simply wrong.

Although behaviorists have been part of the criminal justice system for years—*criminology,* as opposed to *criminalistics,* is a behavioral science—this avenue of research grew steadily in the 1960s and 1970s. This period was both the heyday of social psychology and the time of greatest funding for research on these questions, much of it coming from the federal government through the now-defunct Law Enforcement Assistance Administration.

While the behaviorists had primacy in the 1980s, the 1990s have seen the reintroduction of the theory of biological or possibly genetic determinism into the discussion regarding violence and crime. In some ways, the 1990s have become the decade of the serial killer. Americans are fascinated by the extensive media coverage of the gruesome exploits of such killers as Andrew Cunanan, a homosexual hustler who murdered at least four men, including the fashion designer Gianni Versace; Jeffrey Dahmer, who murdered and gruesomely disposed of more than a dozen young men and teenage boys in Milwaukee; and Joel Rifkin, the Long Island, New York, landscaper who was stopped by police for driving his pickup truck with no rear license plate while he was transporting the body of 1 of the 17 young women he later admitted to having killed. In Russia, "Citizen Ch" was convicted of murdering 53 young people.

Perhaps the decade's icon of gore was a character in Thomas Harris's novel *Silence of the Lambs,* the "tailor" who was killing young women and skinning them in order to make himself a "girl suit" to wear. He was apprehended by an FBI BSU trainee, who in turn was assisted by another serial killer, the evil Hannibal Lecter.

The 1990s also saw the peaking, in 1992 and 1993, then the beginning of a gradual falloff in the rate of murders of all types, part of an overall decrease in violent crime as the decade went on. As the century and the decade close, we have been again debating the causes of crime and violence.

THE "CRIMINAL TYPE"

The debates about whether a "criminal type" exists go back more than a century. Beginning in the 1860s, soon after the first writings by Darwin about the origins of species, Cesare Lambroso, an Italian professor of psychiatry, turned his attention to the study of criminality. Lambroso, who often is called the father of criminology, had helped the police to capture a highway robber, Vallella; after the criminal was executed, Lambroso studied his skull. He discovered a dent on the inside of one of the bones in Vallella's neck and found similar marks through external examination on the necks of many prisoners. Lambroso concluded from this that habitual criminals are "atavistic," a throwback to man's ape ancestors. They could be identified, Lambroso said, by such characteristics as thin upper lips, receding foreheads, protruding ears, long arms, and prominent chins. He also said that they would be shown to have unusual brain structures. Phrenologists of the day, who "read" the bumps on people's heads to determine personality characteristics, also believed that they could determine criminality from the prominence of the region of the head that held destructiveness.

Lambroso also believed that such environmental factors as the person's occupation and poverty could exacerbate criminal tendencies. Thus began the chicken-or-egg questions of criminality: whether criminals are born or made; whether they choose the environment in which they live because of their tendencies or their environment causes their condition; and whether the environmental choice of a parent may doom a child. Out of these questions have come the public-policy debates about how to attack crime: whether to work to catch criminals or to change their living conditions, which presumably cause them to be as they are; whether to punish criminals or to treat them; and whether to change them or help them to change themselves.

In this country, those who make public policy have tried to

distinguish between the criminal and the mentally ill or defective, setting up separate institutions for them and differentiat-
ing between those who are convicted and incarcerated and those
who are deemed unaccountable for their actions and in need of
treatment—although, in many circumstances, the conditions in
which they live while institutionalized are remarkably similar.
In the early part of the twentieth century, in many states, there
were eugenic movements that sought to forcibly sterilize both
criminals and the mentally defective.

In the 1960s, more complex notions of genetics led to a new
idea of biological determinism in criminality, especially violent
crime. The thesis—since discredited—was that men with an XYY
chromosome pattern, as opposed to the typical male XY pattern,
were to be found in great numbers among the ranks of violent
criminals. Because many of the studies were done entirely on
prison populations, the evidence was dismissed—it was impossible to determine whether the XYY condition appeared any
more frequently in these populations than in the general population. Although this turned out to be a dead end, researchers
have determined that hormonal abnormalities do play a part in
deviant—often criminal—behavior; for instance, Depo-Provera,
a synthetic form of the female hormone progesterone, is used in
the treatment of sexually deviant men.

With today's more powerful observational technology that
allows scientists to explore the body and brain in great detail, the
arguments over heredity versus environment in criminality are
as hot as ever, and the notion that body chemistry and nutrition
may be strong contributing factors in criminality is being
researched extensively. Neurobiologists, biochemists, and forensic psychiatrists and psychologists are looking at the human
internal biological and biochemical environment for answers to
the questions of criminality.

In the past, the argument over whether a particular individual was criminally insane or merely criminal was based purely

on psychological and psychiatric assessment. Today, however, neurology—the brain's organic composition—is becoming an added dimension of such assessments. Brain-imaging techniques such as positron-emission tomography (PET) scanning allows neurologists and psychiatrists to explore the brain itself and compare the brains of criminals or criminal defendants with the brains of normal subjects. In PET scanning, an individual is given a radioactive sugar solution that is absorbed by the brain—the brain transforms sugar into energy. The subject is then asked to lie on her or his back with the head in a scanning device and is given a series of tasks to perform. Researchers watch where in the brain the radioactive solution travels; the regions that show the greatest concentration of the solution are those areas of the brain that work the hardest at each particular task.

PET scanning was originally used to detect brain tumors, but scientists are increasingly using it to determine whether there are characteristic abnormalities in the brains of individuals with such conditions as chronic depression or drug addiction, as well as criminal behavior. Some common characteristics have been identified in the brains of addicts and alcoholics, as well as in the brains of children of alcoholics. This has led some to theorize that there is an "addictive brain."

The attempt to link biology and genetics to violence—indeed the entire attempt to "medicalize" violence, makes a number of people edgy. In 1991, then President George Bush unveiled the Federal Violence Initiative, an attempt to treat violence as a public-health issue, including attempts to identify youths at risk for becoming violent and to create appropriate interventions. Scientific studies to identify possible biological markers, nutritional deficiencies, brain-chemistry abnormalities, and other identifiers of youths at risk of becoming violent were suddenly hot items, and National Institutes of Health (NIH) money flowed into the subject.

Then Secretary of Health and Human Services (HHS) Louis

Sullivan liked the idea, and a major conference, "Genetic Factors in Crime: Findings, Uses and Implications," was scheduled for October 9, 1992, at the University of Maryland. The major funding for many of the research projects that were to be discussed at the conference came from the NIH, especially from the Alcohol, Drug Addiction and Mental Health Administration (ADAMHA) and the National Institute of Mental Health (NIMH).

After an indiscreet remark by Frederick Goodwin, ADAMHA director and a major authority on manic–depressive illness, however, the conference was scuttled on political shoals. When discussing the Violence Initiative before the National Mental Health Advisory Council, Goodwin found himself on a tangent about hyperaggressive monkeys, where males kill males in their pursuit of females to copulate with. In what is surely one of the most indiscreet statements by a scientific authority figure, Goodwin said, "Maybe it isn't just a careless use of the word when people call certain areas of certain cities jungles."

Critics of the initiative, and of all efforts to link crime and biology, jumped on Goodwin's remarks as racist and showing the clear racist quality of the initiative's focus. HHS Secretary Sullivan, himself an African American, dismissed the notion that scientific study of biological connections to violence is inherently racist any more than sociological study of violence is. Nevertheless, he had to back down and cancel the conference.

Despite the end of the Violence Initiative in that form, the National Research Council did issue a 464-page study of the state of violence research in November 1992. In that report, the council devoted only 14 pages to biological explanations.

"It's questionable whether genes would have much effect on crime among people who live in an American city," said Gregory Carey, a behavior geneticist at the University of Colorado at the time. "There's no DNA segment that codes for crime. There's a multitude of social, personality, cognitive, and other variables that interceded between the level of DNA and a complex behavior like crime. Carey's review of studies that have tried to link heredity to crime was published as part of the National Research Council's report.

Since 1992, a few studies have continued to focus on other biological or biochemical bases for violence. A set of three articles published in 1994 in the *Archives of General Psychiatry* pointed to several neurotransmitters, including serotonin, that seem to play a role in aggression and dangerous behavior. Also, researchers in the Netherlands identified a rare hereditary disease in which afflicted men were impulsive and violent.

In 1992, a study of 800 boys in public school in Pittsburgh showed that those with higher levels of lead in their bones were more prone to aggressive and delinquent acts than their peers with lower lead levels in their bones. None of the lead levels were high enough to cause lead poisoning, however. In critiquing the lead study, a good question to ask was: How much of this seemingly "biological" difference can be attributed to environment? In other words, were the boys more aggressive and delinquent because they had a higher lead level; or was the higher lead level indicative of the fact that the boys were more likely to live in poverty (in homes where lead paint had not been abated) and to suffer from the emotional and physical strains of poverty?

This question of causality is also asked of the brain-scan studies showing that violent men share some brain abnormalities, especially those cased by trauma. Is the brain abnormality the cause of the violence, or did the conditions under which these men lived—violence and abuse as children—cause their violent behavior and lead to the brain trauma and subsequent abnormality on the scan?

Once biological conditions such as serotonin levels, lead levels, or brain abnormalities have been found, is there any way to abate those conditions? Should children with low serotonin levels—which may, over time, lead to increased impulsiveness, violence, depression, and addiction—be medicated early in life?

Should those violent adults found to have brain abnormalities be confined for life because their brain scans show them to be at increased risk for becoming or continuing to be violent offenders? These are tough questions—questions that science can only help us ask. Public policy must answer them.

SERIAL MURDER

According to the FBI, until the 1960s, nearly 80 percent of all murders could be attributed to friends or acquaintances of the victim, and the motive was usually greed or passion (spouses killing spouses, friends or relatives killing each other, drug dealers and other criminals killing each other over turf or business deals gone bad, planned and paid-for executions carried out by organized crime). These murders often were easy to solve—often, there were witnesses, or at least people who knew of the tension that existed between the victim and others. Today, the number of stranger murders—where the victim and killer have never before met, has more than doubled. Often, these murders are committed during another crime, such as a robbery, but in many instances— some estimates say as many as 5,000 each year—the murders appear motiveless, even bizarre, the work of a seriously deranged person. Each year, thousands of unidentified bodies turn up, many of whom police believe to be the victims of murder.

Statistics show that between one third and two thirds of these people probably were victims of *serial killers* (those who kill one individual at a time over a period of time). The total number of serial-murder victims can be estimated by adding the number of people known to have been murdered by serial killers and the number of missing persons assumed to have been murdered by serial killers. Using these calculations, Ronald Holmes and James De Burger, in their 1988 book *Serial Murder,* estimate that between 3,500 and 5,000 people each year are killed in the United States by serial murderers. The vast majority of serial-killing victims are women and children; nearly all serial killers are men. The killings tend to be intraracial: Blacks kill blacks, and whites kill whites.

Given that known serial murderers usually have been shown to kill between 10 and 15 people per year throughout their violent careers, Holmes and De Burger estimate that there are about 350 serial murderers in the United States at any one time.

The FBI profilers have developed a system that classifies serial murderers and sexually motivated murderers (so-called lust murderers) into two categories: organized murderers and disorganized murderers. Each type of murderer behaves in certain ways while committing the crime; these murders are classified by how the crime scene appears.

Disorganized crime scenes suggest that the murderer did not mean to kill the victim; a sudden rage or fear of being remembered might have driven the killer, or the killer may have attacked a random victim in a blind frenzy.

On the other hand, an *organized crime scene* suggests that the killing was planned and that the killer chose the time and place. Often, the crime scene will be where the body was found, but not where the victim was killed, as in the New Bedford slayings.

The FBI profilers study crime-scene photographs and the reports of investigators who searched the crime scene for the most minute details. They also study the medical examiner's report on the cause and manner of death and the weapons used; whether there was postmortem (after-death) activity such as hair cutting, evisceration, or deliberate mutilation of sexual organs (often seen in killings by immature, sexually naïve killers, who tend to be disorganized); whether there is evidence of torture before death (often seen in killings by organized killers who have a psychological need to punish); and whether the victim was involved in sexual acts shortly before—or even after—death.

The profiler needs to get a complete investigator's report, including a presumed reconstruction of what happened, detailed witness interviews, and a detailed time line of the victim's travels and activities immediately before the killing. The profiler also needs background information on the victim's age, gender, race, and physical description, including dress when last seen and when the body was discovered. The profiler needs to know details about the victim's marital status and adjustment, demeanor, sexual adjustment, personality, intelligence and achievement, occupation, personality and lifestyle characteristics, location of the last few residences in relation to the crime scene, physical and mental medical history, use of alcohol or other drugs, and friends and enemies. All this information can help to determine common characteristics of victims in a series of killings, or if there were any characteristics about the victim of a seemingly random, motiveless killing that might tell something about the killer.

From this information, the profiler tries to determine the characteristics of the murderer, rapist, or molester: age, gender, race, size, marital status, and adjustment; sexual adjustment; socioeconomic and employment status; and even some personal characteristics about demeanor and condition of the subject's home, car, or personal attire.

For instance, says Alan Burgess, a special agent in the BSU, "You may see that the victim was carried from room to room in a house or apartment. There may be evidence that the killer was looking for something. He may have washed up, or changed clothes. This all indicates that the person feels a certain amount of confidence in staying in that place after he has killed someone. It may indicate familiarity with the person or the place" and may suggest that the victim was chosen deliberately rather than at random; the killer may have stalked his victim.

If the victim's face is covered, and she is moved to a place where she will be found easily, it may signal a killer who knew his victim. He may have fantasized about her, found her unapproachable, and, after getting up the courage to speak with her, been rejected; he may have killed her in a blind rage or even inadvertently. "He may have a speech impediment, or a physical deformity, or acne," says Burgess. "He may not go after a peer. He may be very socially inadequate. He might be older but approach very young victims. This kind of person will often act out only very close to home, because he feels safe there."

In addition, by finding live victims who can describe the criminal's behavior at the crime scene during interviews after the fact, FBI profilers and researchers have been able to draw a picture of the conditions that may have led to the behavior.

Not all sexual murders are the work of serial killers, but the work of serial killers is overwhelmingly sexual. Disorganized crime scenes, especially, often are the work of one-time killers or killers who, because of their especially peculiar behavior, are destined to be caught. As Burgess says, "The more bizarre the crime is, the more specific you can be in the profile. There is so much behavior there that the range of people who could have done it is narrow."

A crime scene left by a disorganized killer often will look sloppy. It will be clear that the victim was killed at the scene and the body left in the position in which it was when the murder was committed, with no attempt to conceal it. There will be evidence of sudden violence to the victim, suggesting that the killing itself was spontaneous. The murder weapon often will be a weapon of convenience rather than one the killer brought with him, and it often will be left at the scene. The killer often depersonalizes his victim: There is no attempt to re-dress the body after a sexual assault, and there is evidence of overkill, such as multiple stab wounds or repeated beatings by a blunt object. There often is evidence of sexual acts that occurred after death— evisceration or mutilation of genitals or breasts that implies "experimentation" or "curiosity" about anatomy—especially female anatomy—rather than any ritual or attempt to take a "keepsake" from the victim.

The disorganized killer (often called an asocial killer) usually is below average in intelligence, socially and sexually immature and inadequate, and unskilled. He lives alone or with parents or siblings into adulthood. He often is one of the youngest children in his family and comes from a family where the father's work was unstable or where the father was absent, either physically or emotionally, because of a weak personality or due to alcohol or other drug abuse. He often was the recipient of harsh discipline as a child, from either a violent father or a dominant mother, and he almost certainly has had association with dominant females—his mother, a sister, a girlfriend, or a wife.

Disorganized killers are moved to kill by situational stress and are anxious and troubled during the crime. Many undergo significant behavioral changes after a killing, such as drug (e.g., alcohol) abuse or hyperreligiosity. They usually choose their victims randomly—although, because they often live and/or work near the crime scene, they may know their victims by name. They are not highly mobile and feel comfortable only in narrow geographic areas.

Disorganized killers usually carry out a fast attack on the victim, often from behind. They often kill immediately to get control of the situation, then overkill the victim. They may obliterate the face of an individual they know or who represents someone who causes them distress.

Far more dangerous is the methodical, organized killer. When psychiatrists examine these people, or even just find out the details of the crimes, they often label these people as schizophrenic or paranoid, but more often as sociopathic—meaning that they have no sense of responsibility to anyone or anything in society and no sense of remorse for activities that they know, intellectually, are deviant.

Holmes and De Burger have classified organized serial murderers into four groups:

1. *Visionary Type.* These killers are led to commit murder by visions and voices, often "edicts from God" or from "demons" who speak to them and command them to murder. An example of this type of killer might be the Son of Sam, David Berkowitz, who said he received messages to kill women and often killed them by shooting them through the windows of cars in which they were necking with their boyfriends.

2. *Mission-Oriented Type.* These killers commit murder to rid the world of "undesirable" people, such as prostitutes. They are not necessarily commanded by any vision to carry out the mission, and they often understand that there may be consequences to their actions. Nonetheless, their self-appointed role as a defender of decency or desirability is so strong that it leads them to kill. The New Bedford killer might be a mission-type serial killer, as might the Green River killer in King County, Washington, a still unknown person to whom more than 40 killings of prostitutes have been attributed. The same can be said of Joel Rifkin, who described the 17 women he said he killed from 1991 to 1993 as prostitutes, although police did not believe that all of them were. Perhaps the most famous mission-type serial killer was Jack the Ripper, who killed prostitutes in nineteenth-century London and left messages for the police telling them that the women needed to be destroyed. (One might be tempted, on cursory examination, to categorize Jack the Ripper as a disorganized killer because he eviscerated his victims; however, he was a very skillful cutter, and speculation always has been that he was a surgeon.

3. *Hedonistic Type.* Subcategories of this type are the "thrill-seeking" killer, for whom the actual act of killing produces a high similar to drinking alcohol, taking other

drugs, or driving a car fast; the killer for whom killing helps to enhance the comforts of life, such as a spousal serial killer, who murders a series of husbands or wives for their money; and the lust murderer, who gets sexual satisfaction from killing or from a sexual experience—usually rape—that either inevitably or accidentally turns into a killing. Theodore Bundy, who, before his execution in Florida in January 1989, admitted to over 30 murders in Washington, Oregon, Utah, Colorado, and Florida and possibly killed as many as 300 women, would fall into this category. (Bundy was once asked by investigators if it was true that he had killed 36 women; the story is told that his answer was "add a one to that." No one knew whether he meant 37, 136, or 361.)

4. *Power/Control-Oriented Type.*   This is the killer who feels the need to torment and torture people and gets ultimate satisfaction—not necessarily sexual—from having total control over the life and death of a person. Bundy also showed signs of falling into this category: An ex-girlfriend spoke of her relationship with him being one in which he needed to be in command, tormented her psychologically, and forced her into painful sexual experimentation, although from what is known about the murders he committed, he often killed in a swift, brutal fashion.

Organized killers plan their crimes. The victim usually is a stranger, although she or he is targeted for some reason. For example, Wayne Williams, the Atlanta serial killer, targeted black teenage boys and young men who almost always were somewhat small in stature—under 150 pounds. Ted Bundy's victims—at least while he was choosing individual victims—were pretty young women with long hair parted in the middle, who often looked like a woman he had dated and wanted to marry.

The crime scenes left by organized killers often show that they had overall control of the situation, personalized the victim, and demanded submission by the victim; the killers often leave rope, handcuffs, or other restraints that they used or the victims show evidence of having been bound. There often is evidence of violence to the victim before death, possibly even sustained torture. Any weapons used in the murder usually have been taken from the crime scene. There often is evidence that the victim was transported to the crime scene, either before or after death. If the crime scene is a "dump" site, it is close to a road, where the body is sure to be found, but is not the scene of the murder. If the murder site ever is discovered, it usually turns out to be secluded or a residence, either the victim's or the killer's.

The profile of the organized killer (often called an antisocial or nonsocial killer) is that he is average or above average in intelligence and socially and sexually competent and skilled, although he may work at jobs below his skill level. Ted Bundy, for instance, dropped out of law school in Washington, possibly because he feared that he was becoming a suspect in the series of murders that he was committing around the university where he was a law student. He enrolled in law school in Utah, but dropped out again shortly before being arrested. Wayne Williams was a photographer. Robert Bardella, who pleaded guilty in 1988 to murder and may have killed a dozen young men in Kansas City, was an artist and owner of a shop that sold beads, jewelry, and other art. Many organized killers are living with a partner at the time they commit their crimes: Bundy had a longtime live-in girlfriend, and Albert DeSalvo, the Boston Strangler, who claimed to have raped, molested, peeped at, or harassed hundreds of women over a period of 15 years or more, was married and had two children.

The organized killer often is among the oldest children in his family and comes from a family where the father's work was steady. Discipline during childhood often was inconsistent, and,

Burgess says, "He usually took the most abuse from his father," although he also often "had a background of prior association with a domineering female." These killers usually are "highly intelligent and highly mobile," Burgess says, with a car in good working order. There often is situational stress that precipitates the murder. They may use alcohol or other drugs in conjunction with the crime, but they maintain a controlled mood.

Many organized killers follow the news-media coverage of the crime and may keep newspaper clippings or even take "remembrances" from each crime. They often are interviewed in conjunction with the crime, and some even actively inject themselves into the police activities surrounding the crime, showing up at crime scenes that are being investigated and being "helpful" during their interviews. Burgess says that these killers often inject themselves more as they kill more, and the killing becomes a game between them and police.

(In perhaps the greatest irony, Bundy was able to keep up on the investigation of the missing young women in Seattle in the early 1970s because of his friendship with Ann Rule, a writer of "true-crime" articles for a number of magazines, with whom he worked as a volunteer at the Seattle Crisis Center. Rule's book about Bundy, *The Stranger Beside Me,* was a best-seller.)

Police often limit information about a killing to weed out false confessions. In a serial murder case, it is especially important that some information be kept from the press and the public, so that the killer does not become more careful to avoid leaving such clues, as Wayne Williams did when he started stripping his victims to their undershorts, in an attempt to keep them from being found with incriminating fiber evidence.

"In our research talking to serial killers, assassins, and rapists, we've learned that these guys learn from their mistakes," Burgess says. "They often modify their methods, and may even stop murdering for a while." Although there usually is no outward behavioral change after the crime, these killers often move on after killing a number of times, leaving town and relocating, only to commit another series of murders in a new town later. "These guys may start getting into trouble with the law when they are 15 or 16," Burgess says. "They kill for the first time in their early 20s, and are about as good as they are going to get [at killing] by the time they are 25."

Because the organized killer is competent socially and sexually, he may pick up his victim, rather than abducting her or him, and may develop a pseudorelationship with the victim. The killer may approach the victim in a bar, at a convention, or on a college campus; most of Bundy's victims were college students. He may approach the victim asking for help, to call a tow truck for a broken-down car, for instance.

Bundy often used the ruse of wearing a cast on an arm or a leg and asking for help in moving something or changing a tire. Police who searched his various apartments, as well as people who knew him, often saw plaster of paris and gauze in his home. On Sunday, July 14, 1974, a man calling himself Ted, with a cast on his arm and his arm in a sling, approached at least eight women at Lake Sammamish State Park near Seattle, asking them for help in getting a sailboat onto his car. Two of those women, Janice Ott and Denise Naslund, were never again seen alive; the other six women told police of "Ted's" request when they read or saw news accounts of the missing women.

Rape, rather than murder, may be the conscious motive, and the organized killer may have raped before he started killing. Some life trauma, a situation where a former rape victim almost identified him, or behavior by the victim that he was not expecting could trigger his first murder.

The organized killer may have elaborate fantasies about the crime, and the two activities, the fantasy and the actual killing, may become circular—the killing may lead to better fantasies that then lead to more elaborate planning for the next attack. Many organized murderers also carry out elaborate rituals in their attacks, and investigators may clearly see the crimes as the work of the same individual. Albert DeSalvo, who murdered his 11 victims in and around Boston from 1962 through 1964 by strangling them with their own clothing—usually nylon stockings or the sash from a robe or housecoat—tied the garrote in an elaborate bow around his victim's neck and also often tied another bow around one of her ankles. Some organized killers ejaculate on their victims or on their victims' clothing, or mutilate their victims in a particular way, often cutting off a body part such as an ear, a finger, or a nipple as a keepsake.

Often, the murder weapon is not found at an organized crime scene, suggesting that it is a weapon of choice and not convenience, used repeatedly by the killer. In an organized dump-site crime scene, the victim often will be placed in a specific place or posture. DeSalvo left his victims with their clothing in disarray and their legs spread wide, exposing their genitals, usually pointing at the door to their apartments.

Crime scenes may appear mixed in terms of whether a killer is organized or disorganized. For example, David Berkowitz—the Son of Sam—chose victims deliberately: women parked in cars with men. His background, however, clearly labeled him as a disorganized offender; he was a postal service employee, lived alone, and had a history of institutionalization for mental illness.

Sometimes, experts in serial murder point out, a murderer gradually will move from organized to disorganized crimes; he may kill more frequently and more recklessly. In the early years of his known murdering period, Ted Bundy carried out his murders methodically. In Seattle, he left a number of women in one location in a park, although they were abducted from a large geographic area. In Colorado, he preyed on women on ski vacations. Those who knew him in Washington, Utah, and Colorado remembered a good-looking, charming college and law student who didn't have trouble getting a date and was articulate and even "desirable" in many ways, although manipulating and conniving. In Florida, however, where he eventually was arrested and convicted and ultimately was executed, he carried out a rampage through a sorority house, killing and maiming women, then was arrested for abducting and murdering a 12-year-old girl.

Often, the killings get closer together, or a string of killings will end abruptly when the killer is arrested on an unrelated charge. Ted Bundy was arrested in Utah for possession of burglary tools and was convicted in a kidnapping, but he never was charged with any murders in that state. Bundy never admitted to any murders until a few days before his execution. Albert DeSalvo, who had been arrested and served a brief jail term in 1961 for molesting women by posing as a photographer's agent, knocking on women's doors and taking their measurements, was arrested for breaking into a woman's apartment, tying her up, and fondling her; he did not rape her or brutalize her. However, he confessed almost immediately to being the Boston Strangler.

In addition to their detailed examination of crime-scene evidence and investigative reports, the FBI profilers also bring to their job a detailed knowledge of statistical studies of convicted rapists and killers in sexually initiated murders, which have shown that the majority are young and have trouble both in relationships with women and in holding a job. Therefore, saying that a killer was in his mid-20s, was divorced or single, and had difficulty holding a job is not going very far out on a limb. Neither is saying that the killer is white when the victim is white, nor that he likes to torment people when a ligature is found around the victims' necks, as in the Long case in Florida.

THE MURDERING MIND: OUTLOOK OR ILLNESS

Research conducted by FBI BSU personnel, working in conjunction with academic researchers, suggests that a rich fantasy life also is an important component in the makeup of those who commit the most brutal sexual murders. They argue that serial and sexual murderers are motivated by an outlook rather than by a clinical mental illness. This outlook is the product of psychological stresses imposed on them by family interaction and upbringing. Not coincidentally, these researchers believe that whatever "treatment" is given to those who possess the murdering mind must be carried out in a punitive prison setting, rather than in a clinical hospital setting. Serial murderers are not insane, these researchers argue, but they are possibly evil.

Others, however, argue that to do something as horrific as committing a series of planned executions of innocent, random individuals in and of itself is evidence of a mental illness. While the individual may not fit the legal definition of *insanity*—being unable to understand the charges against him and/or unable to assist attorneys in defending him against those charges—the individual is certainly "ill" and not "evil," and deserves to be treated that way.

In fact, today a number of states are trying to have it both ways. They try serial rapists and murderers as criminals, yet after the individuals have served the penal sentences, they try to have them recommitted to mental health facilities as "sexually dangerous predators."

As a country, we are moving away from both the notion of confinement for the purpose of treatment and imprisonment for the purpose of rehabilitation, to a methodology of confinement in whatever place possible, for as long as possible, simply for retribution and to keep undesirables off the streets.

In an effort to determine which argument is more credible—the wrongdoer with an evil outlook needs to be punished or the mentally ill perpetrator needs to be treated—investigators have turned to research.

The first project to investigate the "outlook" argument and the first systematic study of sexual homicide from a law-enforcement perspective rather than a clinical one, was a study of 36 men convicted of sexual murder. They were all first- or secondborn, were relatively intelligent, and were relatively physically attractive adults. Most came from two-parent homes with fathers who worked and mothers who were homemakers. So: What went wrong, and when?

The study, published as "Sexual Homicide Patterns and Motives," and with significant portions published in the FBI's *Law Enforcement Bulletin,* showed that in most cases, there were a number of factors, intrinsic and extrinsic, in common:

- Half of the men had family members with a history of criminality; over half had family members with a history of mental illness.

- There was alcohol abuse in 70 percent of the families, other drug abuse in one third.

- There was "family instability"; only one third grew up in the same location and one sixth moved frequently; over 40 percent lived outside the home before they were 18, in foster homes, state homes, detention centers, or mental hospitals.

- Half of the fathers of these men where physically absent from home for some time before the boy was 12, and 60 percent said that the dominant parent in their formative years was the mother, regardless of whether their father was present.

- Nearly 70 percent reported "uncaring" relationships with their fathers, and nearly 50 percent reported such relationships with their mothers.

- Forty percent of the men who responded said they were victims of physical abuse as boys, 70 percent said they were victims of psychological abuse, and 40 percent said they were victims of sexual abuse.

- Their sexual interests were solitary in nature; pornography, compulsive masturbation, fetishism, and voyeurism ranked high on most lists of sexual interests.

- Despite high intelligence, the men were almost uniformly underachievers in school, in employment, in the military, and in personal and sexual relationships. Many repeated grades in elementary school, and most left school before finishing high school.

These men reported a strong fantasy component from early in life, suggesting that they sought comfort and refuge from poor family relationships in self-stimulus. Said one: "I knew long before I started killing that I was going to be killing, that it was going to end up like that. The fantasies were too strong. They were going on for too long and were too elaborate."

Often, these fantasies turned violent. The worldview of these men took on a coloration in which they devalued people; they described themselves as "loners" with no social attachments from an early age. One recalled that "I used to do my sister's dolls that way when I was a kid, just yanked the head off her Barbie dolls."

They are self-centered. They view the world as unjust toward them and often blame other people for their low performance and lack of ability to fit in and adjust to social situations. They view authority as inconsistent and unstable. Their desire is to be strong, powerful, and in control. Often, this desire becomes an obsession and is manifested through violence. Their sexuality is autoerotic and violent. To these men, in effect, fantasy is reality.

Their deviant behaviors—rape, murder, mutilation, and torture—have roots in both their background and their outlook on life. These researchers, and others who have conducted subsequent studies, argue that these criminals are made, not born, and that the vicious cycle of family dysfunction, rather than genetics, dooms future generations to criminality.

This sounds much like the argument of sociological "liberals," who argue that poverty, homelessness, broken families, and other factors in the lives of the "underclass" are mostly responsible for crime. Indeed, it seems as though the traditional "right" and "left" on the argument may have joined each other as to the causes, but they still differ as to the way to treat the ills—lock 'em up, or change their lives.

PROFILING OF CRIMINALS

The notion of profiling is not new. During World War II, the Office of Strategic Services (OSS), the forerunner of the CIA, asked a psychiatrist, William Langer, to create a profile of Adolf Hitler in an attempt to determine how he would behave in the weeks and months when it appeared that his defeat was inevitable. Langer's book *The Mind of Adolf Hitler* was published in 1972.

In 1956, the New York City police asked James A. Brussel, a Greenwich Village psychiatrist who had consulted to the police before, to help with the case of the so-called Mad Bomber, who had been setting off bombs in public places seemingly at random for 14 years and taunting police with letters. Brussel told police that they probably were seeking a rather ordinary-looking man, quiet, polite, proper, and well-dressed. He was between 40 and 50 years old, foreign born—probably central European—a Roman Catholic, and fairly well educated. He was sexually abnormal, probably single, lived with his mother or maiden sister in Westchester County or Connecticut and, when police found him, he would be wearing a double-breasted suit, buttoned.

George Metesky, who was arrested soon after in Waterbury, Connecticut, fit this description perfectly, down to the buttoned double-breasted jacket. Over the years, Brussel worked many times with police, and many of his cases are described in his 1968 book, *Casebook of a Crime Psychiatrist,* in which he told how he had created the Metesky profile. The crime of bombing obviously was a crime of a paranoid personality, and paranoia tends to reach its peak after age 40. Bombs are a common form of protest in Europe, and the grammar in the Mad Bomber letters suggested eastern or central Europe. Most central Europeans are Roman Catholics, and the largest population lived at that time in Westchester County and southern Connecticut. He was sexually abnormal, and most likely not married, because the bombs were shaped like penises and the W's in his letters had rounded bottoms—like cartoon figures of women's breasts—despite the fact that all the other letters were printed in block print. Having that personality, and being part of a community that cherished family ties, he probably lived with family members. The bomber was meticulous and feminine—possibly homosexual—and, Brussel suggested, the neatest, primmest, and most protective male attire was a buttoned double-breasted suit; after all, the bomber knew that the police were after him.

Brussel's comment about the W's in Metesky's letters may sound silly, almost a caricature of Freudian analysis. Within the FBI BSU and in university psychology departments, however, research has been done into psycholinguistic profiling—creating a psychological profile based on the language used in threatening letters. In addition, handwriting experts say that the individuality of handwriting mirrors the individuality of people in general. Lee Waggoner, an FBI special agent and handwriting analyst, wrote in the FBI's *Law Enforcement Bulletin,*

> The basic premise that no two people write exactly alike is a generally accepted tenet within the community of document examiners and has been accepted by the courts. The physical act of writing is a habit or a reflexive action. As a person writes, his mind is on the words or message he is trying to communicate. Therefore, in normal writing the movement of the muscles required to push the pen across the page, forming the letters and words, is controlled by the subconscious, while the conscious mind concentrates on the message.

Brussel helped police to write the bomber a letter, which was published in the New York *Journal-American*. The bomber replied and told of his need for revenge against a utility company. A search of Consolidated Edison records led police to Metesky, a 53-year-old, unmarried, former Con Ed worker living with two unmarried sisters.

Although Brussel was so successful in the Mad Bomber case, he unwittingly helped to lead police astray in the Boston Strangler case. In late 1963, Massachusetts Assistant Attorney General John Bottomly, who had been put in charge of a statewide investigation by the attorney general, created a medical–psychiatric committee of experts headed by Donald Kenefick of Boston University's School of Legal Medicine, to try to create a profile of the Strangler. Although the psychiatrists on the panel were familiar with what Kenefick called the "general profile" of a sex murderer—a man with a rage against an important figure in his life (usually a dominant female), who engages in powerful, sadistic fantasies in which he kills that figure—many of them still imagined that such a twisted, demonic mind would have to live in a shambles of a body. Intellectually, however, they knew that such a man might be outwardly normal, even exceptionally pleasant, polite, neat, and punctual; these are traits often found among confidence men, who are sociopaths.

Brussel suggested first, that the killer was of southern European heritage because garroting is a popular method of killing there, and second, that he lived in the throes of a violent Oedipal complex, killing mostly older women who looked like his mother; this did not, however, account for the five young victims. Brussel had an answer for that, too, though. Part of the man's psychological struggle was with his lack of potency; most of the women had been molested with objects, and there was no evidence of intercourse or even ejaculation, except in the last killing, of 19-year-old Mary Sullivan, in which the killer ejaculated on the carpet next to her body. Brussel suggested that the killings had stopped in January 1964, six months before the medical–psychiatric committee met, because the man had, in a gruesome way, "cured" himself.

When DeSalvo was arrested, and he confessed, it became clear that he did not in any way fit the profile the committee had worked up. He had not come from a home with a domineering mother. He was not a homosexual—police, before the profile was constructed, had spent thousands of hours investigating in Boston's homosexual community. He had been heterosexually active since his early teens, had had sex with older women as a teenager, and was "oversexed"—his wife told police that he demanded sex five or six times a day. He had been honorably discharged from the military—hardly an emotional basket case—and had worked steadily in construction for ten years.

Brussel's profiling work, like that done for police for many years by most other consulting psychiatrists and psychologists, was done from a perspective of psychiatric diagnosis and treatment. The subjects of profiles often were described in clinical psychiatric terms, such as borderline personality or paranoid schizophrenic. The problem with this format is that it does not translate very well from the clinician to the investigator; the concept of a psychopathic monster is very different from the idea of a criminal whose behaviors, while clearly deviant, often are ordered and rational within his own context. The shifting of focus in profiling, from describing a condition of mental illness to describing a set of behavioral characteristics, allowed profiling to become far more helpful to investigators.

As Burgess puts it, "Where we broke from psychiatrists and psychologists is looking at things from a law enforcement point of view, not looking for a label, a diagnosis. A policeman could not care less if this person is treated, he wants to know 'how can I get the evidence and how can I lock him up? How can I approach this person who I know is highly intelligent and wants to play a game with me? How can I bring this person back into the investigation?'"

PREEMPTIVE PROFILING

Unfortunately, this nonclinical type of profiling, which depends on facts and statistics, can be perverted when carried out by people who are not trained in criminal investigation and who are not trying to create a profile in order to match a specific individual to a specific incident. In the past quarter century, it has led to situations where "preemptive profiling" has been attempted.

One attempt at preemptive criminal profiling was made in the 1970s by psychologists working with the Federal Aviation Administration (FAA). In the wake of the sudden spate of skyjackings during those years, psychologists reviewed the circumstances of every past skyjacking to find characteristics common to the skyjackers. Whether it was the profile or other security measures that led to authorities' increased ability to deal with skyjacking is impossible to tell, however. (The FAA has never released details of the profile.) However, the creation of security choke points, through which everyone boarding a plane theoretically must pass, and the legality of relatively random searches of passengers who match the profile, are far different circumstances from those of a manhunt for a murderer in an open area that might be a city, a county, or even the entire country.

The FBI, CIA, and State Department have done extensive research on profiling potential terrorists, and the Nuclear Regulatory Agency has spent time trying to determine what the specific nuclear terrorist might look, sound, and act like. During a 21-year career with the federal government, Jerrold Post (currently a professor of psychiatry, political psychology, and international affairs at George Washington University) founded and led the Center for the Analysis of Personality and Political Behavior. The Rand Corporation also does extensive contract research for the federal government in the areas of terrorism and personality.

One profile, the Drug Enforcement Administration's (DEAs)

so-called "drug-courier profile," not only has not been successful at stopping drug smuggling, but has caused a rash of court cases on the grounds that it is overly vague. In the 1970s and 1980s, the DEA developed a profile of the typical drug smuggler, a profile that was so vague and generalized that the net it cast was challenged vigorously by civil libertarians. Among the characteristics that agents look for when matching individuals to the drug-courier profile are:

- Youth
- Luggage without identification tags, empty luggage, or lack of luggage
- Making a phone call after deplaning
- Unusual nervousness
- Use of public transportation
- Last-minute arrival or deplaning last
- Paying cash for an airline ticket
- Appearance of Hispanic origin (Can a police officer really decide what nationality a person is by his or her looks?)
- Purchase of a one-way ticket
- Arrival from a known drug-import center

There are other characteristics that the DEA will not talk about, nor will it reveal exactly how many items of this set of characteristics must be apparent for a "match" to be made. The Supreme Court, in the 1979 case *Reid v. Georgia,* ruled the DEA drug-courier profile unconstitutional, calling it "a somewhat informal compilation of characteristics believed to be typical of persons unlawfully carrying narcotics." However, in another case, *United States v. Mendenhall,* the court had ruled the profile constitutional under some circumstances. In the latest decision, in 1989, in *United States v. Sokolow,* the court ruled 7 to 2 that when taking the "totality of the circumstances," drug agents are allowed to detain a person who fits the drug-courier profile and search the person and his or her belongings.

In 1984, DEA agents stopped 25-year-old Andrew Sokolow on his return to Hawaii from Miami, where he had stayed for just two days. A drug-sniffing dog found cocaine in Sokolow's shoulder bag. Sokolow, wearing a black jumpsuit and gold jewelry, had paid $2,100 in cash from a roll of $20 bills for the airline ticket. The court ruled that the DEA did not need probable cause to detain and search Sokolow, only "reasonable suspicion" for the brief "investigative stop." (Once, when I was returning from Malaga, Spain, in November 1981, with no luggage because of a baggage-handling problem in Madrid, I was stopped at Kennedy Airport and subjected to a search because I fit the DEA profile.)

Chief Justice William Rehnquist, writing for the majority, wrote that nothing Sokolow did was "by itself proof of any illegal conduct," but that the DEA nevertheless had the right to stop him. "While a trip from Honolulu to Miami, standing alone, is not a cause for any sort of suspicion, here there was more: surely few residents of Honolulu travel from that city for 20 hours to spend 48 hours in Miami during the month of July. . . . Any one of these factors is not by itself proof of any illegal conduct and is quite consistent with innocent travel. But we think taken together they amount to reasonable suspicion."

Justices Thurgood Marshall and William Brennan, in their dissent, however, argued that "reflexive reliance on a profile of drug courier characteristics runs a far greater risk than does ordinary, case-by-case police work, of subjecting innocent individuals to unwarranted police harassment and detention."

According to a 1986 study by the Congressional Office of Technology Assessment, at least 16 federal agencies have used computers to generate statistical profiles of individuals who may commit crimes, most notably fraud against various federal programs.

RAPE-VICTIM INTERVIEWS AS SOURCES OF CLUES TO RAPIST PERSONALITY

In rape cases where the victim survives, investigators and pro-filers have the advantage of working with an individual who can testify to the interactions she had with her assailant. Too often, however, police who interview rape victims do so ineptly; they sometimes try to be too tactful and neglect to get from the victim telling detail because it is too embarrassing or because the victim wishes not to discuss it. By concentrating on trying to get the victim to remember what the rapist looked like, the investigator often heads up a blind alley. If the interview with the rape victim is thorough and thoughtful, however, and it gets from the victim the details of the entire interaction—physical, verbal, and sexual/psychological—the profiler can gain an understanding of the rapist that can't be obtained from investigative work or from thorough examination of crime-scene photographs in killings. Also, the profilers know that in many cases, today's rapist is tomorrow's sexual killer.

Rapists reveal themselves to profilers through three kinds of behavior: physical, verbal, and sexual. While rape is a sexual crime, its motivation often is not sexual but is part of a pattern of dominating, aggressive behavior that is motivated by both intrinsic and extrinsic forces acting on the perpetrator. When profiling a rapist, the profiler needs to know, in the most minute detail possible, the rapist's behavior in all three areas.

Rapists approach victims in much the same way killers do: indirectly through a "con" or subterfuge, or directly by a "blitz" or surprise attack meant to subdue the victim. Once the rapist has control of the situation, he may maintain that control through mere presence, verbal threats, displaying a weapon, or actual force.

The level of intimidation and force used during the rape can help the profiler to determine the rapist's motivation. Victims may not distinguish among the physical, verbal, and sexual assaults, but the interviewer and profiler break apart these three activities to discover often subtle clues to the offender's motivation and, ultimately, his identity. Rapists may use minimal force—no physical force to mild slapping; moderate force, such as repeated slapping or punching, accompanied by profanity; excessive force, including severe beating, profanity, and personal degradation; or brutal force, including sadistic torture that may even lead to death, coupled with profanity and personalized verbal abuse.

The profiler also needs to know what level of resistance, if any, the victim put up—passive, verbal, or physical—and if the offender escalated his level of force in response to the resistance. Profilers have found five different rapist responses to resistance: ceasing the demand, compromising, fleeing, threatening, or escalating the level of force.

Other areas that the victim interview must cover are any sexual dysfunction the rapist suffered, the type and sequence of sex acts that occurred during the rape, and the verbal activity of the rapist and the victim. These activities can range from rapists making victims ask for sex or say "I love you" to either the rapist or the victim engaging in extremely graphic language; from a rapist who wants to watch a victim disrobe to one who tears her clothes off; from kissing and fondling to brutal intercourse. These activities all tell the profiler whether the rapist is motivated by a need for love or by rage and hatred and whether he is punishing and dehumanizing his victim.

Although the terms *organized* and *disorganized* are not necessarily used by rape profilers, parallels between the ways that profilers analyze the behavior of the rapist and the killer are apparent.

The victim also can help investigators to tell whether the rapist is new to the activity or a veteran, by recalling what measures he took not to be identified, how much cleaning up he did after the fact to remove trace and physical evidence of the incident, and whether he thought through his escape route before or during the early phases of the incident. Did he do such things as search out an escape route; disable the telephone; bring weapons, a gag, and restraints; have the victim wash after the attack or wash items he had come into contact with during the attack?

What, if anything, the rapist takes from his victim also can give information about the rapist. He can take evidence such as those items he touched or those that may have his semen stains on them, which would suggest that he has experience and knows about incriminating evidence. He can take valuable items, which may imply that he needs to rob because he is not employed. He may take personal items, such as a driver's license, lingerie, or a photo of the victim, to remind him of the offense and help him in fantasizing about his next attack.

Arson

Arson investigators also have had success using psychological profiles. While the casual observer often will attribute a rash of fires to a "pyromaniac," experts believe that such motiveless arson is rare. Arson almost invariably is motivated by one of five things:

1. Organized crime, such as loan sharking, extortion, and concealment of other crimes.

2. Insurance fraud

3. Commercial fraud, such as inventory depletion or a desire to modernize

4. Residential fraud, such as a desire to relocate, redecorate, abandon an old automobile, or get public-housing authorities to do something about conditions

5. Psychological needs

Psychological motivations for setting fires include acting out aggression, hostility, or revenge; gaining attention; acting on delusions, hallucinations, or "visions"; sexual gratification; and thrill seeking. Statistically, the psychological arsonist is a white teenage boy with below-average intelligence and poor academic and social performance, who comes from a home environment that is disruptive, harsh, and unstable, and where the father is absent and the mother is dominant.

Postpubescent and adult arsonists have poor relations with women, lack social skills, have difficulty holding jobs, and often have physical deformities. Arsonists often have been diagnosed at a young age as having psychiatric disturbances, often are alcohol abusers, and often have aggressive sadistic tendencies. It might be said that the arsonist is a disorganized murderer who takes out his aggression on property instead of on people.

There also appears to be an arsonist analog to the organized murderer; this individual often is seen as the "would-be hero" arsonist, the volunteer firefighter arsonist, or the fire-buff arsonist. This type of arsonist often can be seen around fires, having been alerted to them by his police/fire scanner radio, and he seeks to attract attention to himself as being a hero. He may start the fire so that he can call it in to authorities, be the first to respond to the scene, or rescue someone and be a hero. These people often are more intelligent than other arsonists, but they are socially inept, immature, and inadequate underachievers.

In 1988, a rash of fires in southern New Hampshire, which caused a lot of property damage but injured no one, was discovered to have been the work of would-be heroes, men who worked for the volunteer fire departments. In another type of would-be hero incident, in New Jersey, a volunteer firefighter and a volunteer ambulance rescue worker were charged with six counts of attempted murder for allegedly dropping concrete blocks from a highway overpass onto cars, then rushing to the rescue. Similarly, a Long Island hospital worker was convicted in 1990 of murder in the deaths of people he had injected with a drug that caused cardiac arrest, in an effort to revive them and be a hero.

Speaking of the New Jersey case, Harvey Schlossberge, the former director of psychological services for the New York City police department, told *The New York Times,* "These people have a Superman fantasy life, with themselves as the heroic rescuer. But they can't be a rescuer unless there's someone in trouble. So they create their fantasy in order to act it out."

Mort Bart, another psychologist/consultant to the New York city police, said of the New Jersey would-be rescuers, "Their lives are enriched by their fantasies. The harm they do is overwhelmed in their own minds by the good they do. The harm is so subservient to their psychological needs that the criminal nature of their acts is lost to them."

BOMBINGS

It took months to work out the deal, but finally, on January 21, 1998, Theodore Kaczynski admitted that he was the Unabomber. He agreed to plead guilty to four federal counts of murder and admitted to a total of a dozen and a half bombings over a nearly 20-year campaign of mail-bomb terror. Kaczynski's plea bargain —prosecutors agreed to a life-without-parole sentence instead of a trial where they would ask for the death penalty—came at the end of a tortuous process of determining both Kaczynski's competence to stand trial and his general mental health.

For months, Kaczynski had argued with his court-appointed federal public defenders, who wanted to defend him on the grounds of "diminished capacity," an argument saying in effect that an individual has such severe mental illness that he or she cannot have his or her actions judged by the normal standards of behavior. Kaczynski, the reclusive former mathematics professor, refused to allow his attorneys to make that claim, repeatedly telling the federal judge overseeing the trial that he would demand new lawyers if his attorneys insisted on putting forth any evidence of mental illness or mental incapacity. He also refused to undergo psychiatric evaluation by prosecution or defense experts.

In mid-November, 1997, on the eve of the trail, *The New York Times* featured a long takeout article about the Unabomber trial, under the headline, UNABOMB TRIAL TO EXPLORE SANITY AND RESPONSIBILITY. The judge had ruled that neither side could put forth formal evidence by experts about Kaczynski's mental status, although he did say that the defense could present mental-status evidence at the penalty phase of the trial if a jury found Kaczynski guilty in the guilt phase. He also said the defense could question nonexpert witnesses such as family members about Kaczynski's behavior over the years, in a way as to imply deteriorating mental health.

Just before the trial was to begin, Kaczynski asked the judge for yet one more closed-door meeting, again to complain about his lawyers' plans for his defense. Kaczynski asked the judge to allow him to defend himself, or to hire an attorney who would argue that he was a political prisoner. In a three-day delay, during which the judge was trying to iron out these difficulties, Kaczynski apparently tried to commit suicide.

At this point, the judge ordered a psychiatric evaluation by a federal prison consulting psychiatrist, and he ordered Kaczynski to cooperate or to be transferred to a federal prison hospital for an involuntary 60-day psychiatric evaluation. The psychiatrist, Sally Johnson, most famous for conducting the psychiatric evaluation of John Hinkley after his unsuccessful assassination attempt on President Ronald Reagan in 1981, found Kaczynski to be competent to stand trial despite suffering from "schizophrenia, paranoid type."

This finding—that an individual can suffer from one of the most debilitating forms of mental illness, one that calls for long-term if not lifelong medication and often for long-term periods of hospitalization, yet still be competent to stand trail for a capital offense—strikes many people as odd.

This paradox is possible because the legal definition of competence—or insanity for that matter—has little if anything to do with a medical definition of mental illness. Also, courts have long given states the right to move that line to one place or another in any way to suit the states' needs. That's why prosecutors can get convictions of serial rapists despite defense efforts to have them deemed not guilty by reason of mental defect, only to have the state turn around years later, after the individual has served his sentence, and argue that he should be confined to a mental hospital after he has served his prison sentence because he is a sexual predator and has a mental defect.

In the Kaczynski case, his plea agreement said nothing about where he would be confined, in prison or in a mental-health facility; nor did it force the government to treat him for his mental illness. By finding him competent to participate in a legal proceeding and to be judged in a legal proceeding, did the court and the court-appointed psychiatrist deny him the rights any of us have to be treated for our illness?

7

ELECTRONIC SURVEILLANCE: IS BIG BROTHER WATCHING AND LISTENING?

In the movie *The Conversation,* audio surveillance specialist Harry Caul (Gene Hackman) has a crisis of conscience because the information he provides to his client could get someone killed. Beyond the moral question, *The Conversation* provides a chilling picture of how vulnerable our most intimate discussions can be. When the film was released in 1974, on the heels of the Watergate break-in and revelations about governmental surveillance of President Richard Nixon's "enemies," the movie was one more reminder that with increasing technological sophistication, privacy is ever more threatened.

In *The Conversation,* Caul is asked to listen in on a conversation between two young people in a public park in San Francisco. He and two associates use three different microphones—one stationary and two mobile—to pick up snippets of the conversation. The stationary device is a parabolic microphone

that can be rotated to follow the action, and the mobile microphones are mobile only because they are being carried around by
two people wandering through the crowd, trying to stay near the
people being eavesdropped on. Later, in his workshop, by electronically enhancing the three recordings and removing much of
the background noise, Caul is able to put together a composite
and listen to the entire conversation.

Caul's array of electronics is a far cry from the simple telephone tap that has been used—legally and illegally by both
police and private citizens—for much of the twentieth century.
At the same time, it is only the tip of the electronic-surveillance
iceberg. In a 1988 report on surveillance technology, the U.S.
Senate Judiciary Committee's subcommittee on constitutional
rights created an array of five categories of electronic surveillance
that were possible, although not all were in use at that time
because the technology was not widespread in the general public. They are

1. Audio surveillance
    a. Miniature transmitters that can be carried around
    b. Wired devices such as telephone taps and
        concealed microphones
    c. Tape recorders

2. Optical/imaging surveillance
    a. Photographic techniques
    b. Closed-circuit and cable television
    c. Night-vision devices
    d. Satellite-based surveillance

3. Computer-based surveillance
    a. Microcomputers
    b. Distributed processing networks
    c. Software such as expert systems
    d. Pattern-recognition systems

4. Sensor technologies
   a. Magnetic sensors
   b. Seismic sensors
   c. Strain sensors
   d. Infrared sensors
   e. Electromagnetic sensors

5. Other surveillance devices and technologies
   a. Citizens-band (CB) radios
   b. Pen registers (devices that register the telephone numbers called)
   c. Vehicle-location systems
   d. Machine-readable magnetic strips
   e. Polygraphs
   f. Voice-stress analyzers
   g. Laser intercepts
   h. Cellular telephones

In the decade following the 1988 report, this array of technology expanded to include digital fingerprint technology, palm-print identification by cash machines, recognition of an individual's eye (used in the 1998 Winter Olympics in Japan), pens that recognize signatures, and all kinds of "smart" tracking badges that tell where in buildings visitors or employees have been and whether those wearing the badges have washed their hands after using the toilet.

By using a combination of these technologies, one can monitor movements, actions, communications, and even emotions. No doubt, these technologies have a legitimate place in policing. However, as Gary Marx, a sociology professor at the Massachusetts Institute of Technology, argues in his 1988 book, *Undercover: Police Surveillance in America,* these technologies also allow law-enforcement authorities to undertake wide-ranging "fishing expeditions" and to place the United States at risk of becoming a "maximum-security society."

In preparing its 1985 report, "Electronic Surveillance and Civil Liberties," the Congressional Office of Technology Assessment polled 142 divisions or components of federal agencies to find out which were using electronic surveillance and what kinds were being used. Thirty-five governmental components reported some use. (The CIA, the National Security Agency [NSA], and the Defense Intelligence Agency [DIA] were excluded from the request, on national-security grounds, although it is well known that these agencies use significant amounts of electronic data gathering.)

In this poll, the three users of the most diverse array of technologies were the Drug Enforcement Administration with 10 different technologies; the FBI, with 9 technologies; and the Customs Service, with 9 technologies. These agencies also reported plans to use a lot more surveillance technology by the 1990s; the FBI said that within five years, it would be using 17 different technologies, the DEA said that it would be using 11, and the Customs Service said that it would be using 10.

In fact, during the 1990s, these and other U.S. agencies did add a lot of surveillance technology to their repertoire. While official reporting was reduced during the 1990s, due to budget cuts that decreased the amount of communication out of the government about its current and planned uses of these technologies, more nongovernmental watchdog agencies sprang up to keep an eye on governmental use of surveillance technology.

In addition, much technology that was prohibitively expensive until the late 1980s became cheap enough so that local and state police forces are using it, too. Also, with the demise of the Cold War, some surveillance technology that used to be reserved for the military and highly classified police operations are now available to civilians. Even more contemporary military technology is entering the civilian market. Sophisticated tracking and locating technology used in the Gulf War of 1991 (to pinpoint Iraqi troop movements) are now advertised by car companies, to aid in personal navigation. You can buy a dashboard system that uses location triangulation, locates your car, and suggests the best route to the destination you key in. For as little as $300, any individual can purchase a precision photograph from a civilian satellite that rivals the NSA satellite photography.

In 1985, the most popular forms of electronic surveillance were closed-circuit television, with 25 agency components using it and 4 more planning to use it by the 1990s; night-vision systems, 21 using and 1 more planning to use it; and miniature transmitters, 19 using and 2 more planning. No one reported using expert systems (discussed in Chapter 8) or voice recognition, although 3 agency components were planning the use of each technology. One agency component was using satellite-based visual surveillance, and one was intercepting microwaves; one more planned to undertake each. One agency even reported that it was planning to intercept fiber-optic communications.

With all this effort—and money—being spent by agencies of the federal government, many people share Gary Marx's concern: "Once the new surveillance systems become institutionalized and taken for granted in a democratic society, they can be used for harmful ends," against those with the "'wrong' political beliefs, against racial, ethnic or religious minorities, and against those with life-styles that offend the majority." In short, these technologies hold within them the roots of totalitarianism.

Legal Protections against Surveillance

The protection of citizens from overzealous law-enforcement surveillance has come through a patchwork of Supreme Court decisions and federal legislation. Most of this legislation and case law is based on the Fourth Amendment right of citizens to "be secure in their persons, houses, papers, and effects, against unreasonable searches and seizures." With each new technology, police and prosecutors once again test this framework. Despite this constant challenge, however, the basic message of privacy as it is conceived in legal terms is this: The concept of "privacy" in this society, derived from protections in the Fourth, Fifth, Tenth, and Fourteenth Amendments, protects a person, not necessarily a place (there is no specific right to privacy in the Constitution). In communication between people, there is a "reasonable expectation" of privacy and, therefore, it is necessary for law enforcement to get a warrant to intercept that communication. In much the same way, a person should be free from warrantless search of personal correspondence and work documents, even if they do not fall under the traditional definition of "papers."

The Electronic Communications Privacy Act of 1986 (ECPA) tried to take into account all new forms of communication and to define which ones deserve privacy protection under the Fourth Amendment and Title III of the 1968 Omnibus Crime Control and Safe Streets Act, and which do not.

THE TECHNOLOGIES OF COMMUNICATION AND OF COMMUNICATION SURVEILLANCE

Telephone

Until recently, a telephone was easy to define—an analog device that transmitted oral communication by wire from one party to another. Today, however, telephones are much more complex. Despite this increased complexity, the technology necessary to eavesdrop is still relatively simple. Although evidence derived from such wiretapping is inadmissible if there is no warrant for the tap, there is no doubt that some amount of unwarranted telephone eavesdropping is done—by law enforcement, under the guise of criminal intelligence, and by the country's intelligence agencies. For years, the FBI tapped the telephones of political dissidents, and Henry Kissinger, while secretary of state, is known to have tapped the phones of some of his staff workers.

Increasingly, the analog voice signal that goes in one end of a telephone is digitized. Further, not only voice, but also data and image are increasingly transmitted via telephone. In addition, in many instances, only part of a telephone communication is transmitted by wire, the rest going through fiber-optic cables, microwave radio transmission, and satellite links. In many countries with a history of poor telephone service, the 1990s have seen an explosion of cellular networks, which are—as we approach the twenty-first century—almost exclusively digital.

It is possible to eavesdrop on conversations as they travel by microwave or satellite without physically tapping a phone line. The NSA, responsible for the country's "signals intelligence," plucks literally millions of individual conversations out of the air daily.

There are at least two instances where the NSA apparently picked up microwave and satellite communications that turned out to have possible law-enforcement ramifications (if one defines foreign terrorism against U.S. citizens as a U.S. law-enforcement issue, or as a law-enforcement issue of the country where the terrorism took place, as well as a national security issue). Apparently, it was an agent of the NSA who heard Egyptian President Hosni Mubarak telling other Arab leaders that he had got rid of the hijackers of the cruise ship *Achillie Lauro* by putting them on an airplane to North Africa. U.S. Air Force pilots forced the plane to land, and the hijackers were arrested and tried for killing a U.S. passenger on the ship. In another instance, the NSA apparently picked up telephone communications from the Libyan People's Bureau in East Germany, telling officials back home that it had just blown up a discotheque in West Berlin. In this case, German officials were concerned with the law-enforcement aspects of the case, while the U.S. government used this information as a rationale for bombing Libya shortly thereafter.

A digitized telephone conversation is no more difficult to intercept than is an analog conversation. Encryption devices are available for those who want to defeat eavesdroppers for either type of device. Rather than using a simple connection on the main telephone wire leading from a certain line that picks up the sound pulses, the digital-telephone tapper needs only a coder/decoder (similar to a computer modem) and some knowledge of the modulation scheme—the particular rate at which data "bursts" are sent.

Cellular telephones work by transmitting a radio signal from the phone to a base station in the geographic cell from which the call is placed. The base station then uses the same technology to route the call to another base station (if the call is to another mobile cellular phone) or to a telephone system (if the call is to someone using a conventional wire telephone). Cellular systems can transmit data, as well as voice. Tappers of cellular systems have only to find the proper radio frequency to intercept a call between the caller and the base station, between base station and base station, or between the base station and the point at which the call enters the telephone wire system.

Mobile radio telephones are cheaper than cellular phones. Although they have limited range, a customer can subscribe to a service that provides access to "repeaters" to get the signal over longer distances. Again, an eavesdropper needs only to find the radio signal to listen in. During the 1996 political campaign, a Florida couple "inadvertently" intercepted a conference call among Republican Congressmen and political consultants and gave a tape of the call to a Democratic congressman. The ensuing controversy focused both on the content of the discussion and the interception.

Cordless phones provide no privacy at all, and the Federal Communications Commission (FCC) requires a warning on the phones that the signal as it travels from the cordless phone to the individual's base telephone can be overheard accidentally, even by a neighbor listening to his or her FM radio. The same is true of the infant monitors sold to parents; stories abound of parents hearing a screaming child and finding their child asleep soundly, only to realize that the child on the monitor is the child down the street.

Since telephone deregulation, no longer are so-called common carriers the sole source of telephone service. Many companies have set up private branch exchange (PBX) systems, which are as easy to eavesdrop on as any other telephone system. A PBX allows employees of a company to speak with anyone in any company office in the country, possibly in the world, on an internal telephone system, rather than calling on the system of a common carrier such as AT&T, MCI or Sprint. Many government agencies also use PBX systems, which sometimes are called tie-line systems. The capital cost of a PBX is great, but the company no longer has long-distance charges for calls within the organization.

Transactional Information

It is possible to learn a lot about a person's communication without actually listening in. Transactional and billing information is stored in *real time*—meaning that the information is accumulated as it is happening—and can be obtained through such technologies as pen registers, automatic billing equipment, and so-called trap-and-trace mechanisms.

A *pen register* senses changes in magnetic energy that correspond to the numbers dialed by a telephone caller. With a rotary phone, a very sensitive radio receiver can pick up these pulses. Because touch-tone phones produce a much weaker magnetic pulse, it is necessary to put a device on the wire itself to pick up the signal. Pen registers, which must be attached with the consent of the phone company, can determine not only the number called, but also the length of the call. The trap-and-trace device allows one to work "backward"; it is placed on a receiving telephone to determine the number from which a call originated.

Computer-controlled electronic switching has eliminated much of the need for a pen register. The switch controller automatically collects information on the call to determine whether it is a flat-rate or a toll call. This information is retained, and with the phone company's permission, access can be gained to the computer tapes that store the records. A simple reverse telephone directory for the geographic area where the receiving telephone prefix is located is used to determine to whom the calls were made, whether a pen register or electronic transactional information is used. The same is true for finding the number of the sending telephone when using a trap-and-trace mechanism. Contemporary caller ID systems allow anyone to retain a record of the phone numbers from which calls have been made to them.

Electronic Mail Surveillance

Although an increasing number of people are zapping their communications rather than putting them in envelopes, most are oblivious to the possibilities for easy eavesdropping offered by electronic mail. From the old standbys of telegraph and telex to Fax machines and voice mail, electronic mail has no specific privacy protection by law; before the ECPA was passed in 1986, however, many experts believed that an individual could claim a right against government interception under the Fourth Amendment.

There are five stages at which transmission of electronic mail is especially vulnerable to interception. The first stage is when the communication is in the sender's terminal or the computer system's memory. Anyone, be it a law-enforcement official or a "hacker," can gain access to the information and read it or alter it, merely by correctly dialing the access code into the system. This can be done either by gaining knowledge of the access code or by using a device that continuously dials new combinations of numbers in an effort to find the correct access code. After that, it is a matter of breaking the password security that insulates each user in an electronic-mail system. Most computer experts say that if messages within the system are not encrypted, it is not very difficult to read other people's electronic mail.

The second stage is in transmission. The technique for interception here is the same as for a voice communication: tapping into the wire or fiber-optic cable, or intercepting the microwave or radio transmission.

In the third stage, interception can be made in the electronic mailbox of the receiver, whether the mail is in the receiver's computer terminal or in an electronic mailbox in a central computer network, and whether that mailbox is in a computer system at the receiver's location or in a public mailbox system that the receiver has rented.

The fourth point at which interception can take place is when the mail is printed into hard copy before mailing. Although once the message is in an envelope for the receiver, it has the same protection as a letter, it is vulnerable during the time it is being printed out.

Finally, if retained by the electronic-mail company, either in hard-copy form or on computer tape or disk, the information is vulnerable to interception.

Database and electronic communication privacy has been an increasingly hot issue through the 1990s, as more and more businesses and households become wired through services such as America Online, as well as simple Internet access providers. At the close of 1997, a number of Internet access providers held a major symposium on the subject of Internet privacy.

The major topic, even more important to individuals than the security of their private conversations via E-mail, is the degree to which personal information about an individual is available. In a report prepared for the Federal Trade Commission by the FBI, there was a sixfold increase from 1991 to 1997 in reports by individuals of providers allowing wrongful access to information about them stored in computers.

Increasingly criminals are stealing the Social Security numbers of individuals from computer databases and reselling them for the production of false identification or credit cards, or they are merely using the numbers to apply for new credit cards, which are then "maxed out." Individuals then find out that people using their names—and their Social Security numbers—are perpetuating frauds, thereby subjecting them not only to ruined credit ratings, but also to civil and criminal penalties.

Paging Devices

At one time, a person wearing a paging device on his or her belt could be assumed to be a doctor. By the mid-1980s, however, these little electronic boxes could be seen dangling from the waists of plumbers, computer-repair people, public-relations executives, and drug dealers. Today, teenagers and parents often wear pagers, to notify each other of their whereabouts at any given moment in increasingly hectic families. One pager company advertises its product on television with a woman executive reading on her alphanumeric pager the message, "Hi, Mom, I'm home doing homework."

Pagers are increasingly vital to criminal activities, as well. In almost any raid that police carry out on a drug operation, they now find not only narcotics, cash, and guns, but also electronic pagers. Drug runners, often teenagers, are "beeped" by their bosses when they are needed to make a delivery, and by 1988, many school systems around the county were barring pagers on school premises in the same way they bar guns, knives, and brass knuckles.

Paging devices come in three chief varieties: tone only, voice, and digital display of a phone number or of a message spoken into the phone after activating the pager. All three use radio transmission to send the message from the paging service to the receiver.

- A *tone pager* beeps or hums (they often are referred to as beepers in this country and bleepers in England) to alert the receiver that the paging service has a message for him or her. The receiver must telephone the service to get the message.

- A *voice pager* allows a sender to call a phone number and, rather than leaving a message with an answering service, to use the paging-service facility to send the spoken message directly to the person wearing the pager.

- A *digital-display pager* allows the sender, after calling the paging service, to dial a coded message that is transmitted to the person wearing the pager.

Under Title III, only the voice pager is protected from interception; also, it is usually assumed that there is no presumption of privacy under the Fourth Amendment for digital-display and tone-only communications.

ELECTRONIC COMMUNICATIONS PRIVACY ACT [ECPA] PROTECTION

In 1986, Congress broadened the privacy protection held by spoken communication over wire (traditional telephone conversations in the predigital era) to include all electronic communication. By doing so, it explicitly stated that law-enforcement personnel and agencies would need to obtain a warrant to conduct any interception of electronic communication. However, the ECPA has six exemptions allowing warrantless interception:

1. Publicly accessible radio communications, such as AM–FM transmissions; ham radio broadcasts; CB broadcasts; walkie-talkie broadcasts; and marine, aeronautical, or ship-to-shore broadcasts that are not encrypted

2. Tracking devices (called beacons or beepers, and similar to electronic parole monitors) that often are used by law-enforcement personnel to follow the movements of a person or vehicle. (The beacon uses a radio signal, and location can be pinpointed on a screen; these devices often are planted on a person or vehicle by law-enforcement agents, but once they are in place, anyone can pick up the signal by finding the proper frequency.)

3. The radio portion of cordless telephone conversations (but not cellular phones), which have no presumption of privacy

4. Tone-only paging devices

5. Certain types of surreptitious video surveillance (video surveillance is discussed in the next section)

6. Pen registers and trap-and-trace devices

VISUAL SURVEILLANCE

In late 1988, officials in New York city tested remote video surveillance as a method for catching people who ignore red lights. The system, already in use in many European countries, uses sensors embedded in the roadway and cameras mounted on poles. If a driver goes through an intersection against a red light, the car passing over the sensor triggers the camera, which snaps a picture that captures the car's position and its license plate. A summons then is automatically processed and sent to the address of the registered owner.

In 1997, the little town of Lyons, New York, used half of a federal grant of $29,000 to purchase surveillance cameras that can swivel and zoom to be put throughout the town's bucolic downtown area. In the first four weeks of use, the cameras helped police make three drug-related arrests. Scott Forsyth, a lawyer with New York's Civil Liberties Union, said the cameras "smack of invasion of privacy."

The ability to activate cameras remotely makes them infinitely more useful in surveillance. Other invaluable enhancements of cameras for surveillance include such technologies as the "charge-couple," which removes the need for heavy battery packs and allows miniature video cameras to be concealed in such items as briefcases, picture frames, and the lights of department stores and banks. Many of these hidden minicameras can also swivel around 360 degrees, making it virtually impossible to hide from them or sneak up on them and disable them.

Fiber optics allows a camera lens to be put into place in one location, with the actual camera at a remote location. Using this so-called "light-pipe" technology, surveillance can be carried out with only one entry to install the lens; all changing of film can be done off site. In effect, the camera has become like a microphone.

Cameras are not the only visual-surveillance technique that has come of age recently. It is both technically and financially feasible to maintain surveillance on an individual from an airplane thousands of feet in the sky, using the same kind of high-resolution and computer-enhanced visual technology that is found in spy satellites. Night-vision scopes that amplify light up to 85,000 times can be fitted to cameras, goggles, binoculars, or even a telescope to monitor movements more than a mile away. Low-light-level television has been installed in some cities to try to thwart street crime. Infrared television cameras also can be used, detecting infrared radiation and translating it into electrical impulses to produce a black-and-white picture.

AIRPORT SECURITY

Since the mid-1970s, people entering airport concourses have had to pass through metal detectors, and both hand-held and checked luggage have been subject to X-ray photography and random inspection. Even X rays may fail to detect explosives, however, especially those of the "plastic" (really, chemical compound) variety.

Today, new technologies for inspection of passengers, hand-held luggage, and luggage stored in cargo holds promise the possibility of bomb-free airline travel by the early twenty-first century.

The cargo-hold screening device, developed by Science Applications International Corporation under contract to the FAA, works by bombarding baggage with low-energy neutrons, which will be absorbed by atomic nuclei inside each piece of baggage. The neutrons trigger the emission of various gamma-ray characteristics, and measurement of the gamma rays gives a measure of the contents of the baggage, element by element. Such a chemical analysis can detect the presence of any kind of explosive.

This technology cannot be used to check passengers, because it would present a radiation hazard. To check passengers, the FAA has turned to Thermedics Incorporated, which has created a way of "sniffing" vapors to detect bombs. The company has determined that the three major classes of explosives—dynamite, TNT, and plastics—each have a nitrogen–oxygen "signature." The system works by putting a passenger in a small booth and sweeping a current of warm air around the passenger, then pumping the air into a chamber for analysis of the vapors given off. Thermedics also is working on a hand-held detector that would collect a sample of air from each piece of hand-held luggage.

These screening devices are not meant to replace X rays and metal detectors, which detect about 3,000 handguns each year. Security experts see the new devices as the beginning of a multi-modal approach that will, in the future, add other screening devices and sophisticated computer algorithms to integrate the information from the many screenings.

This will, of course, add to both the time it takes passengers to board and the cost of airline travel.

BIOMETRICS AND PERSONAL TRUTH

A number of biometric technologies are either under development or in use, mostly as a way of securing access to sensitive physical areas. These include devices that "read" fingerprints, palm prints, voice, hand geometry, and even the blood-vessel patterns in the retina of the eye.

In *hand-geometry* identification, by measuring such things as the length, curvature, and webbing between the fingers of a hand placed on the reading sensor (shades of Bertillonage), the machine determines whether the person is authorized for a particular access or activity. Already, some nuclear power plants, automatic teller machines, and government installations are using security systems that measure the spatial geometry of the hand for positive identification. The information is stored in a microcomputer on site.

In *retinal blood-vessel* identification, low-level infrared cameras are used to scan the retina of the eye and feed back to a photosensor a picture of the blood-vessel pattern. The information is digitized and computer processed, then stored as a signature template.

Computer analysis of *handwriting* and *voice recognition* may also be used as access security mechanisms. A signature, theoretically, does not change from one day to the next; the computer analyzes such characteristics as speed, pressure, and conformation. While voice recognition has improve measurably in the 1990s, and is currently in use not only in industrial applications but also in desktop computer systems for those who don't or can't use a keyboard, most experts have given up on voice recognition as a security identifier.

Voiceprints have been used in the past for comparison and even have been allowed in court as evidence, although the prints measure electronic pulses of the voice, not the actual voice, and many people believe they are unpredictable.

A multisensor system of personal identification could be developed that would use any combination of these technologies; access to some facilities would be granted only to people whose biometric measurements on a series of sensors are on record as having clearance. The technology of identification has even gone as far as the creation of computer chips that could be implanted in the body and read by scanner. So far, this kind of computer branding has been done only in animals, but the day may come when it is done in humans as well—perhaps in children, the elderly, people with memory problems, or even those on electronic parole (see Chapter 9).

Perhaps more frightening than identification by objective body measurement is the inference of truthfulness or lack of it through measurement of body clues. The polygraph—or so-called lie detector—usually measures three physiological indicators as a subject is put through a series of questions. The indicators are rate and depth of respiration, measured by straps placed around the abdomen and chest; cardiovascular activity as it is reflected in blood pressure, measured by a cuff around the bicep; and the electrodermal response, an indication of perspiration, measured by electrodes on the fingertips.

Although these three indicators show changes brought about by increases in certain kinds of stress, most scientific examinations of polygraphs have come to the conclusion that they are not good at determining whether someone is telling the truth. One reason is that a person's level of anxiety can be changed by the interaction with the polygraph examiner; by the way in which questions are phrased, the speed at which they are put, and the order in which they are presented; and by a host of other factors. In addition, one can train oneself to "beat" a polygraph in the same way that one can train oneself to endure physical or emotional duress and torture. In short, skilled liars can beat polygraphs, producing what scientifically is termed a false negative. At the other end, many truth-telling people who are emotionally charged can be shown by a polygraph to be lying, a false positive.

Although polygraph examinations became increasingly popular during the 1980s, both in preemployment screening and as a random way for federal agencies to test for unauthorized activity, such examinations provide the least accurate results, mainly because the questions asked are open-ended and the person being examined has little or no understanding of the context in which they are being asked. In a criminal investigation, where the subject matter is much more tightly defined and the questioning often is much more focused, polygraph examinations have a higher level of validity.

In 1988, Congress passed the Employee Polygraph Protections Act, which outlawed preemployment polygraph examinations for most private employees—exempting some companies that do sensitive work for the federal government. Since then, some companies have turned to pen-and-paper "honesty tests"—another kind of psychological evaluation—as part of the hiring process, but these have also come under fire. Critics believe that the tests give just as many false positives as do polygraphs and, although test designers say the tests should not be a sole reason for not hiring people, critics believe that that is exactly what is happening. In addition, attorneys with the ACLU's Privacy and Technology Project worry that employers draw inferences about personal lifestyle from the answers to certain test questions.

Honest-test questions come in a number of varieties. There are direct questions that ask such things as whether an employee would report other employees who are dishonest; then there are the more ambiguous questions that ask for answers that can give test scorers clues about a person's work and personal habits.

One problem with honesty tests is a problem inherent in all psychological, sociological, and academic testing: the possibility that the questions are biased by the point of view and value system of the question writers and that whole groups of test takers will flunk the test. In other words, a test written and scored by middle-aged, white, heterosexual men may be taken and flunked by blacks, Spanish-speaking people, women, and gay men.

If the polygraph has scientific shortcomings—the inability to catch true sociopaths because they don't experience the kinds of physiological changes when they lie that other people do—the personal truth tests have sociological shortcomings—the inability to get beyond the values of the test writers and test givers.

Other supposedly truth-determining technologies include sensors that measure voice stress, eye movements, and the chemistry of blood, urine, and saliva. One neurophysiologist at Northwestern University, J. Peter Rosenfeld, has even developed a lie detector that uses electrodes attached to the scalp and a computer to pick up changes in brain-wave activity and display them on a video monitor. Rosenfeld told the Associated Press in July 1988 that "if sensory information has a special meaning for you, your brain will respond with a special kind of trough called the P3 wave." When test subjects are exposed to words that coincide with "antisocial" activities in which they are involved, such as taking drugs or cheating on tests, "their brain waves break into P3 waves," Rosenfeld said.

If the information collected by both public and private sources carrying out such surveillance were used in real time to carry out criminal investigations and enhance the security and public good of those who use public facilities, most of us would agree that they are necessary. When the information is stored and used at a later time or is used for political, social, or religious persecution—as such information has often been used in the past—the implications of putting more powerful technology into Big Brother's hands are increasingly terrifying.

A 1997 American Management Association survey of 906 large companies found that 37 percent of those companies monitor the telephone numbers employees call and the length of time they spend on the phone; 34 percent watch employees randomly with stationary security cameras to guard against vandalism, theft, or violence; 16 percent look into their employees' computer files to measure how many keystrokes they make in a minute or what is on their display at any particular time; 16 percent video-tape performance; and 10 percent tape employees' phone conversations. Denny Lee, speaking for the ACLU's workplace rights program, told the *Washington Post,* which first reported the story, "people shouldn't lose their right to privacy when their workday begins."

Checking up on employees even extends to the toilet. In August of 1997, the Tropicana casino and Resort in Atlantic City, New Jersey, installed Hygiene Guard, computers developed by Net/Tech International Inc. The system of infrared cameras, computers, and tracking software allows employers to check up on employees' washing habits. Unless an employee, who must wear a battery-powered "smart card," at work, makes all the requisite stops after entering the bathroom—the sink for a long enough time, the soap dispenser—the badge will register an "infraction" into the computer storage system. In some instances, the badge will beep at the employee. Some public health advocates applaud the effort, stating that efforts to educate the public and employees about the necessity of washing hands fall short. But civil libertarians are aghast. The ACLU calls the technology "Big Brother in the Bathroom."

Many advocates of increased surveillance argue that by virtue of the mere knowledge that surveillance is being undertaken, far fewer people will engage in illegal or antisocial behavior. Through random monitoring, listening, looking, and even drug testing, these people argue, the society will be more law-abiding and less crime-ridden. On the other hand, putting individuals under constant scrutiny breeds apprehension and fear, and putting the burden of proof on the individual to show that he or she is not engaging in certain activities runs completely against our tradition of civil liberties.

Some critics of surveillance technologies have asserted that totalitarianism inevitably follows from the argument that to be free from social ills, one must give up some freedoms. If the argument is put that way, most U.S. citizens would say "No, thank you" to such technologies. With each new technological development in the world of surveillance, however, social control becomes easier and personal liberties more fragile. Also, when the information gathered by these surveillance tactics can be stored, analyzed, and combined in infinite ways using ever faster computers, the possibilities for loss of freedom multiply in quantum leaps.

8
COMPUTERS
AND COMMUNICATIONS

When the commissioner of the Social Security Administration, Dorcas Hardy, admitted to Congress in April 1989 that her agency had verified 151,000 Social Security numbers the previous year as "a test run" for TRW Inc., the nation's largest credit-reporting company, many people were outraged. TRW had proposed to pay the Social Security Administration $1 million for computer time to verify 140 million Social Security numbers for credit checking purposes. Such a mass verification would be a clear violation of the Federal Privacy Act, which calls for such "confidential" information as Social Security numbers to be released only in cases "compatible with the purpose for which the records were collected."

This reaction may seem surprising, however, at a time when most of us willingly put our Social Security number on dozens, if not hundreds, of documents each year—from job applications to bank accounts to invoices for clients who must report to the Internal Revenue Service the fees they pay to contractors. In fact, the notion that a person's Social Security number is confidential seems almost ludicrous.

While the idea that our Social Security number floats around to virtually anyone who asks for it may or may not bother people, the calculation that—for only $1 million in computer time, salary, and other expenses—the Social Security Administration could verify the Social Security numbers of about 60 percent of the U.S. population attests to just how powerful computers are at amassing, storing, collating, and discharging information.

The power of computers, police and prosecutors argue, is a great boon to law enforcement, but it also can be a great threat to civil and constitutional rights. While the surveillance techniques outlined in Chapter 7 can trap our communications, computer databases can trap the facts of our lives.

Both the technology itself and the people who use it are making the new breed of computer snoopier than ever before. As computers become more complex, they also become more prone to error. Further, the more people who have access to these computers, the greater the chances that human error, political bias, or just plain curiosity will endanger our liberties. When the often disparate facts of one's life are collected and collated, the data become information that can be used to identify, track, and even judge a person.

Imagine the following scenario. You have a relatively common name, James Robinson. It is dark and raining, and you are late getting home for dinner. Near your home, you get to a traffic signal that is yellow, and you speed up to get through before it turns red. A police officer pulls you over and chides you for speeding up at a yellow light on such a miserable night—it's technically illegal, you had enough time to stop, and it's dangerous.

Although he is not inclined to issue you a violation—maybe

a warning—he does take your license and registration back to his car because, with his powerful data terminal, it will take only a couple of seconds to do a thorough search of records, and he has instructions to do a check every time he makes a traffic stop.

On his terminal screen, he sees a warrant from a neighboring state for James Robinson for failure to appear in court on a charge of reckless driving. The description is similar to yours—although the officer can't see much of you in the dark sitting in your car. The warrant carries your Social Security number, although in the state in which you currently live, the Social Security number is not your license number, and the officer has no way of knowing what your Social Security number is. The officer comes up to your car and informs you that unfortunately, he is obliged to place you under arrest on an out-of-state warrant. You are flabbergasted.

At the police station, you think, This must be a case of someone else with my name being wanted. You call your attorney, at home, who tells you she'll be there in half an hour, not to say anything, and everything will get straightened out. Then you sit and wait. You have voluntarily given the police your Social Security number, and, after your attorney speaks to the police, she says that yes, indeed, there is a warrant for you. It is five years old, from the state you previously lived in. The charge is reckless driving.

It takes only a second for the incident to click into focus. You had been leaving the parking lot of your old high school after visiting for a football game. It was Thanksgiving weekend, and everyone was a little rowdy. Some old friends jumped on the hood of your car, and you swerved to get them off, knocking over a trash can. One of the police officers at the game wrote you out a traffic violation.

A date was set for a court hearing, but a little research found that what you had done did not fall under the law the police officer had used; you brought this to the attention of the district court magistrate, who agreed and said that the charge would be dropped.

Somehow, the information that your case had been closed had not gotten to the court clerk who maintained the docket of cases. Your case was called, and because you were not there, a default warrant was issued. You had moved out of state shortly afterward and had never had even a traffic violation until today. The information was still on the central computer of the state you had lived in, however, and when the officer who stopped you queried the nation's police-record computers, the information surfaced.

You tell this story to your attorney, who has you tell it again to the police. You are freed on your promise to appear at a hearing a week later. During that week, your attorney will research the matter and try to get the other state to acknowledge the end of your case and remove the information from its computer.

A week later, the matter is resolved.

THE NATIONAL CRIME INFORMATION CENTER: IN THE BELLY OF THE BEAST

At the center of criminal investigation by computer is the National Crime Information Center (NCIC), the FBI's national database of criminal information records. Just before the turn of the twentieth century, 40 years before the birth of the FBI and 50 years before ENIAC (the first modern computer) the National Chiefs of Police Union urged Congress to finance an agency that would collect and transmit information about "criminals and criminal classes" to police forces throughout the country. For years after the FBI's founding, it performed this task with file cards, folders, letters, telegrams, and wanted posters.

In 1967, the NCIC came into operation, with about 300,000

individual records maintained in five computer files. Today, the NCIC maintains records numbering in the millions upon millions. Data can be entered or retrieved by 64,000 federal, state, and local criminal-justice agencies, using their own local computer systems. Data can even be entered or retrieved by individual police officers using microcomputer terminals in many of today's patrol cars. The NCIC handles nearly 1 million inquiries each day, instantaneously. Critics charge that because of the system's speed, police use it for mass or random record checks, such as keying random license plate numbers into their cruiser terminals or entering the names of everyone on a homeless shelter's sign-in list.

The Interstate Identification Index, known as Triple-I, has accounted for a majority of the records in the total NCIC system since shortly after its inception in 1983. Triple-I is a "pointer" system that answers an inquiry by pointing the inquiring department or officer to criminal information records on file in one of the states that hold their records individually, rather than having the FBI hold them (including most of the larger states). States transmit information directly to the inquiring department after Triple-I notifies the state that a query has been made about a record it holds. Although Triple-I technically decentralizes criminal records, it actually cuts the time necessary to answer inquiries and reduces the FBI's work in maintaining records.

COMPUTER-POWER CRIME ANALYSIS AT THE LOCAL LEVEL

When the NCIC went on-line in 1967, the only way to communicate inquiries was through departmental mainframe computers, which were available only on state or regional levels, or in the largest of cities, because of their price tag of a quarter-million dollars and up.

Today, however, a police department can put a personal computer, with software and printer, in a patrol car for less than the cost of the car. In addition to being allowed instantaneous inquiry through NCIC, patrol officers can now feed their daily logs on-line into police departmental computers. In St. Louis County, Missouri, one of the earliest major users of personal-computer technology, there was a 20 percent increase in officers' on-street time between 1983 and 1985, using rather primitive equipment. By 1989, hundreds of departments had gone on-line, many with even better results.

Today, as the twenty-first century approaches, even three-and four-officer rural departments have computers in police cars. These computers provide them with immediate access to desk-top minicomputers at the station, as do the computers of state police, nearby county sheriff officers, and the highway patrol. In effect, we have a nationally wired police force in this country today, thereby allowing officers to use up-to-the-minute information to analyze crime patterns and adjust the deployment of personnel.

The St. Louis County computer system, known as REJIS (Regional Justice Information Service), was started in 1973 and by 1983 had a mainframe computer, three minicomputers for distributed processing, and 250 terminals. REJIS was used by 84 criminal-justice agencies (police departments, sheriff's departments, prosecutors, courts, correctional institutions, and probation/parole agencies in the city of St. Louis); 92 incorporated communities in the county; communities in 3 other Missouri counties; and 4 counties in southern Illinois, just across the Mississippi River. During the 1980s, REJIS shifted to using more personal computers; those agencies and departments that waited the extra decade to join the computer age didn't need to make this shift, as they started operations on personal computers.

Reporting on the first decade of REJIS, Steven Claggett, the former administrator, noted a reduction in the clerical workload by 30 percent, the production of *automatic arrest notices*—notice of a "hit" when someone wanted in one jurisdiction within the region was arrested in another regional jurisdiction—and automatic printing of summonses, jury notices, and attorney's reminders. In addition to the link with the NCIC, officers could query a number of local databases, including adult arrests within the region; field interview reports, which compiled fragmentary information such as nicknames of people interviewed in the course of investigations, descriptions of them and their vehicles, and their last known addresses; local criminal histories; the dispatchers' logs to reconstruct investigative time lines; businesses that have alarm systems wired into the police and their location, alarm codes, and other information; and files about court and corrections information.

Computerized crime-analysis reports are not difficult programming feats; essentially, data are logged by patrol area or beat, time of day, day of week, type of crime, and any details about the *M.O.* (modus operandi, or method of operation). The computer then can simply break down daily, weekly, monthly, or annual crime statistics by these various factors.

The CARE management system, used in St. Louis, gives division and precinct commanders information about beat deployment, crime patterns, and crime trends. Precinct commanders use this information to determine selective patrol targets for a particular shift or day; because information is on-line immediately after an incident report is filed and can be sorted by reporting district, patrol beat, or even building type, commanders can change patrol strategy in a moment's time.

In New York City, by the end of 1988, 2 out of 75 precincts had their own crime-analysis computer systems in operation—the 13th precinct around Gramercy Park in Manhattan and the 67th precinct in the Brownsville, Crown Heights, Canarsie region in Brooklyn. In the 13th and the 67th, monthly crime reports now are done by computer rather than being hand tabulated, and computer charts have replaced push-pin maps in detailing, street by street, where crimes occurred.

New York really stepped up its computerization efforts in 1994, after the election of Rudolph Giuliani to his first term as mayor. Giuliani, a former United States Attorney for the Southern District of New York (Manhattan), brought in as his first police commissioner William Bratton, former police commissioner in Boston and before that for the New York Transit Authority.

Giuliani and Bratton had three main goals for the police force in New York City:

1. *Streamlining the force.* This was done by combining the city police, the transit authority (TA) police (responsible for the city's subway and bus system), and the housing authority police (responsible for crime fighting in the more than 500,000 units of public housing throughout the city).

2. *Getting more officers on the street.* This was done in two ways: (1) increasing the computerization of much record-keeping and getting officers who formerly performed clerical tasks onto patrol; and (2) hiring more civilians for nonsensitive clerical work, also freeing up officers.

3. *Becoming more active and less reactive in crime fighting.* Giuliani and Bratton had three ways to go about becoming more proactive: (1) beefing up the police intelligence unit, (2) attacking quality-of-life issues on the streets, and (3) establishing sophisticated computerized crime analysis.

They attacked quality-of-life issues by getting rid of public drinking; low-level street-corner drug using and selling; and so-called "squeegee men," (homeless men who wash car windows at traffic lights, whether the vehicle owner wants it done or not, then get aggressive if they are not paid for their services, often vandalizing cars). The mayor and the commissioner thought they could create better relations with neighborhood residents and get more cooperation from them if they addressed these quality-of-life issues.

The two executives also set up a sophisticated computerized crime-analysis system that was to be developed centrally, then fed to all precincts throughout the city. Precinct commanders were to be taught to use the system, and the system's deployment of resources on a daily, weekly, and monthly basis was to be partly driven by "the numbers." Bratton's right hand on this computerization effort was Jack Maple, a brash street-cop turned dandy. Like his boss, Maple relished double-breasted suits and dinners at Elaine's, the ritzy New York restaurant. To his credit, however, Maple was also a wizard at seeing patterns emerge from crime statistics, and at creating techniques for deploying troops to the "hot spots" for better crime fighting.

Bratton and Giuliani often clashed on personal and some political matters, and Bratton left in 1997, but Bratton's successor, Howard Safir, maintained the computerized analysis. Maple also left, forming a consulting company and taking his techniques on the road. Maple's first major project began in 1996 in New Orleans, a police department mired in corruption and poor crime fighting.

Back in New York, Giuliani was reelected handily in 1997, partly on the strength of four straight years of crime reduction in New York City. Some argue that demographics (fewer people in the prime crime-committing ages), more prisons and tougher sentencing, a better economy, and the end of the crack-cocaine epidemic that ravaged the city in the 1980s are the chief reasons for the reduction in crime, but clearly Bratton and Maple's computer analysis of crime patterns also played an important role.

Automated Fingerprint Identification Systems (AFIS): The First Use of Artificial Intelligence (Pattern Recognition)

Automated fingerprint identification systems (AFIS) are the first area in which primitive artificial-intelligence concepts are being used in police work, although many experts in artificial intelligence would dispute whether computerized pattern matching truly could be called artificial intelligence. *Artificial intelligence* is a computer's emulation of human intelligence; the computer software emulates human experience and expertise.

An AFIS has two components. One is the optical reader and digitizing software that can "look at" a fingerprint—either a 10-print card or a latent print—create a spatial geometric "map" of the ridge patterns and minutia of each fingerprint, and translate the fingerprint minutia into binary digital code, which then is entered into the computer's memory. The second component is the mathematical algorithm that allows the computer to search through files and compare fingerprints. The system searches through either a local jurisdiction's file or the FBI Identification Division's fingerprint file containing more than 83 million items. In a matter of minutes, the system compares the new fingerprint with those already in the file and comes up with a group of possible matches, based on a scoring system that assigns points to each of the criteria used by technicians to match fingerprints. The threshold score is set at a point where the computer narrows the field of possible matches—or "hits"—to a reasonable number for a fingerprint examiner to go over in a few hours of work.

Having such a vast search capability allows police to conduct *cold searches* of latent prints, where there are no suspects; this type of search was previously unfeasible because of the masses of fingerprint information on file.

In the early years of AFIS, there were two problems. One was that AFIS systems were often incompatible; there were a few commercially available systems with different algorithms, none of which had software that read prints from other systems. This problem was overcome in 1987, when a national standard was created and software-conversion programs were developed so police departments would not have to scrap AFIS systems bought before the standard was developed.

The second problem was that while AFIS could search large files, it could provide the fingerprint examiner only with an index list of possible matches; the fingerprint cards or latent-print reproductions themselves then had to be manually removed from the card files.

Once a fingerprint—either a latent print or a card—is searched and digitized, it can be stored for later retrieval. So, in 1983, the FBI began working on technology to create high-quality digital fingerprint images. By combining the digital imaging with a second-generation reader, the FBI also has started work on computerized classification of new prints, which will complete the chain and automate the entire fingerprint identification process.

Finally, Fingermatrix, a company in White Plains, New York, has developed a system that scans and digitizes live fingerprints to create computer-image 10-print fingerprint cards, which could eliminate altogether the process of inking and rolling fingerprints at the time of arrest.

Today, the FBI is still working on digitizing and storing all of its millions of formerly card-based fingerprints. Some are now on-line, and the system is available for states and local police departments to query. In addition, many states have moved forward more quickly than the FBI in digitizing their old fingerprint records, and they are putting all new fingerprints on-line in real time. Many can also query each other's systems.

The danger in all this computerization is that as this technology becomes more advanced, it will (as often happens) be carried far beyond law-enforcement use into other areas. Fingerprint identification systems began being used as security for some sensitive corporate and government functions in the late 1980s. Only those persons who have their fingerprints maintained in an authorized user file are allowed to enter, and the system then logs who entered and when.

In the 1990s, a number of states began fingerprinting welfare recipients in an effort to verify them when they came to collect their monthly checks. Many individuals dropped off the welfare rolls. While advocates of fingerprinting said this was probably because they were collecting welfare fraudulently, advocates for the poor argued that no one could be sure the fingerprints were not being used for law-enforcement purposes (police dipping into welfare fingerprints to solve crimes) or that they would not be used for such purposes in the future. If such a system were put into wide use, our movements could be kept track of constantly. Put that together with a more powerful, all-encompassing NCIC, with active tracking files of those under investigation, and the possibilities are frightening.

National Crime Information Center Problems: Past, Present, and Future

The major problem with the NCIC as it now stands is data integrity. The mass of criminal-justice data that must be entered, even in a modest-size police department, often puts a strain on resources. Frequently, the fact that someone has been arrested is entered into the computer within minutes, but the facts that charges were dropped or that the person was found not guilty take days or months to be entered, or are never entered at all.

Since its beginning, the FBI has constantly lobbied for NCIC's expansion. The criminal history file, sometimes called the "national rap sheet," only came into being in the 1980s and was not uniformly welcomed by police. Having the big picture of a person's criminal history may be important to a court probation department in making a sentencing recommendation, but most cops understand as well as any other citizen the danger of making too many investigative judgments based on criminal histories.

One California police supervisor told a congressional researcher that "the idea that a national rap sheet system would make an important contribution to our work is a bunch of baloney. . . . Most of our leads come from citizens reporting a crime. Without these resources, which have nothing to do with computers and criminal histories, we would be dead."

In 1981, the Congressional Office of Technology Assessment (OTA) asked Kenneth C. Louden, a professor at the John Jay School of Criminal Justice in New York City, to assess the round of expansions proposed at that time for the NCIC. Louden interviewed some 140 police officers, district attorneys, and judges, many of whom doubted the utility of the national rap sheet, and a great number of whom complained about data-integrity problems. Louden received access to random criminal-record summaries from three states and the FBI and compared them to the original records found in county courts. He found that

- In North Carolina, only 12.2 percent of the summaries were complete, accurate, and unambiguous to the level mandated by federal law (in California, the rate was 18.9% and in Minnesota, 49.5%). The FBI records, which depend on the original data input from the various state agencies, were only 25.7 percent complete, accurate, and unambiguous. Within the NCIC, the records were about 49.5 percent complete, accurate, and unambiguous up to the federal mandate.

- Of a random sample of arrest warrants on the FBI's 127,000 item "hot list" on one given day, 10.9 percent had been cleared, 4.1 percent showed no record of a warrant at the original site at which it supposedly was issued, and other warrants had numerous clerical problems.

In an effort to mitigate the release of bad data, the FBI works under the so-called "one-year rule": It cannot give out to non–law-enforcement agencies (such as federal agencies, businesses with federal government contracts, or federally chartered banks, which all do preemployment criminal-history checks) any criminal-history information more than one year old without noting a final disposition. The information that is on-line to law enforcement through NCIC, however, has no such rule.

A number of people have sued the police for arresting them using out-of-date, incomplete, or false data, and police officers have even sued NCIC administrators for putting them at risk in civil suits. Courts often have ruled in favor of people arrested falsely, and many data-integrity cases are settled out of court; either result often takes months or years and has a high financial and emotional cost.

In 1987, the MITRE Corporation, a technology design and consulting firm in Bedford, Massachusetts, that works almost exclusively for the federal government, was contracted to design a new, expanded "NCIC 2000" system. MITRE developed 247 recommendations for changes in the NCIC, and the NCIC Advisory Policy Board advanced a number of those to Congress, including proposals to add files to the database of juvenile records, add misdemeanor records, and add DNA patterns.

The advisory board also suggested that the NCIC include information on any individual who is at any time under investigation for a number of reasons, including drug dealing, murder, or kidnapping. Any time an inquiry was made by any participating agency in the country, the agency that added the name to one of these tracking files would be notified. In some instances, the agency making the inquiry would be notified that it had "hit" a subject being tracked; in other instances, a "silent hit" feature would allow the listing agency to contact the agency making the inquiry, but not the other way around.

In analyzing the NCIC 2000 proposals, the House Judiciary

Committee's subcommittee on Civil and Constitutional Rights turned to the Computer Professionals for Social Responsibility, which, in its "Review of NCIC 2000," argued that "the addition of investigative files would dramatically change the NCIC system. It would turn the NCIC from a public record system . . . into a surveillance system. Such files are not authorized by statute and raise serious constitutional issues."

The OTA, in its 1988 special report on "Criminal Justice: New Technologies and the Constitution," concluded that "it is probably impossible for statutory law on privacy and civil liberties to keep up with the rapid development or improvement of surveillance technologies and computer data management technologies."

The proposed tracking lists are not the first of their kind in the NCIC system. From 1971 to 1974, the FBI maintained a secret Stop Index that contained records of more than 4,700 U.S. citizens, many of whom were anti–Vietnam War activists. The practice of using the NCIC to track the movements of people who were wanted not for a crime, but only because of their politics, ran so counter to both our ideas of democracy and the original intent of the NCIC that in the immediate post-Watergate era, the Stop Index was ended.

In 1975, Harold R. Tyler, Jr., deputy attorney general, admitted to Senator John Tunney, the chair of the Senate Judiciary Committee's subcommittee on Constitutional Rights at the time, that the Stop Index's objective was "to enable law enforcement agencies to locate, through NCIC, individuals being sought for law enforcement purposes who did not meet the criteria for inclusion in the NCIC wanted person file."

However, one tracking list has gained congressional approval over the years. In 1983, using the 1981 assassination attempt on President Reagan as justification, the Secret Service persuaded Congress to allow it to use NCIC as an efficient way to keep tabs on a small number of individuals it considers an active threat to its protectees. The Secret Service file contains less than 1 percent of the individuals in the Secret Service's protective intelligence files (28 names in November 1988 and a high of 85 names in 1985). In any inquiry to the NCIC, the Secret Service file is scanned, any "hit" is immediately reported to the Secret Service, and the inquiring department or officer is notified that the inquiry has hit the name of a person the Secret Service deems highly dangerous.

While the Secret Service list is highly selective, and names can be entered only by one agency, lists of people being investigated or watched for alleged drug activity, murder, or kidnapping (especially drugs) could run to hundreds of thousands of names, and the list could be added to by any of the more than a dozen agencies that are active in the country's "war on drugs."

FBI tracking files would be tantamount to a "stop" by a police officer, some legal experts argue, because there would be a permanent record in the NCIC file, showing the time, date, and place of the inquiry. However, that stop could be completely random and, at least, would be likely to be unrelated to the reason the person's name was on the list. If this were so, the stop probably would not meet court criteria for investigative stops, because the officer would have no reason, based on personal observation, to believe that the person was behaving in a suspicious manner. If the officer did know the reason that the person's name was on a particular tracking list, stops merely because the person was on the list could be construed as harassment.

Tracking lists could give police another reason to do massive random searches of the system by entering all names on particular lists in a city, such as all those who slept in a shelter or visited the unemployment office. A corollary problem with tracking lists is that such mass checks, or even individual checks, could be done on the basis of a person or group of people having common characteristics that together create a "profile." Some of these profiles are created through sophisticated artificial-intelligence programs that allow expert investigators to translate their years of knowledge and expertise into a series of "rule bases" for computerized decision making. Others, however, are simply correlations of statistical or demographic information that can be used by anyone with access to a database to group individuals by how they look, where they live, what government programs they participate in, and a number of other criteria.

Beginning January 1, 1998, a number of airlines began using a profile to flag suspected terrorists. The new profile's creation began years earlier, just after the bombing of Pan American Flight 103, which exploded over Scotland in 1990. Some airlines, with the approval of the Federal Aviation Administration (FAA) began experimenting with the profile as early as 1995, and more joined the experiment in 1996 after the explosion of TWA Flight 800 over Long Island Sound in the summer of 1996.

Civil liberties groups and Arab-American organizations have sued to block use of the profile, which they say discriminates against Arab Americans and Arab nationals because it uses such indicators as last name, current destination, previous destinations, and so on to add up "points" for use in the profile, and it leaves too much to the discretion of airline and airport security personnel.

HOLMES MEETS BIG FLOYD

During 1988, Baltimore County detectives spent over 150 hours being interviewed about their investigative techniques in burglary cases in an effort to develop a set of rules by which a computer can help determine possible suspects and help steer the course of an investigation. For instance:

- If the home's occupants were not away for a period of days when the burglary occurred, it probably was the work of professionals.

- Glass cut with a glass cutter is a sign of a professional; a back door ripped off its hinge is a sign of an amateur.

- An organized search of the master bedroom is most often the work of a professional, a missing stereo and television the work of an amateur.

Solving such routine crimes as burglary is one of the most difficult tasks for the police; in Baltimore County, 80 percent of murders are solved, by only 15 percent of burglaries. Will computers help? They may help in training young detectives; they also may, by organizing information and forcing a person who enters data to follow a logical sequence of questions, get better data from uniformed officers who make reports, and give detectives better information to work with. Without suspects, however, it is doubtful that such an expert system will add much to police work.

Artificial-intelligence programs already are in use in a few federal agencies, most notably the Internal Revenue Service and the FBI. At the IRS, at least 10 expert systems are in the works; the first is Raven, a program that helps field agents to determine whether a late-paying taxpayer deserves a waiver of the penalty. The IRS claims that $40 million per yea is lost because of generous or soft-hearted field agents who don't check up on taxpayer excuses for late filings, such as "my house burned down" or "my accountant had a heart attack."

At the FBI, artificial-intelligence programs are being created for investigations of counterterrorism and serial murders; already *Big Floyd* (an artificial-intelligence system) is helping the FBI in labor racketeering cases. In England, the Home Office is working to add a deductive-reasoning component to its *Holmes* program, which is used to manage and analyze the voluminous records in major violent-crime investigations.

Big Floyd was developed by the FBI with assistance form the Institute for Defense Analysis, a major artificial-intelligence research arm within the Defense Department. Big Floyd, named after Floyd Clark, head of the Criminal Identification Division of the FBI, has access to the more than 3 million records of the FBI's Organized Crime Information System.

An investigator can ask Big Floyd to check all information about an individual or an organization against all applicable federal statutes to determine whether there is enough information to charge the individual or organization and, if not, what additional information needs to be obtained. The program may suggest the next investigative step, such as obtaining a warrant for a wire tap. Based on its ability to analyze relational information, Big Floyd may suggest what other individual might have information that could be helpful in charging the subject of the inquiry. In fact, Big Floyd even may suggest ways that investigators may "turn" an individual to provide information about the subject of the inquiry.

The Behavioral Sciences Unit (BSU) is working on a way to get the computer to generate a psychological profile in apparent serial murders or bizarre, so-called stranger murders. Already, the Violent Criminal Apprehension Program (VICAP) computer does a primitive pattern match. It compares and contrasts over 100 M.O. categories of a new case with all other cases stored in its database and provides the VICAP analyst with the 10 best matches—the 10 cases that have the most M.O. similarities to the new case.

In December 1986, David Icove, a systems analyst in the BSU, described the proposed expert system as one that would look at the information provided on the detailed 20-page VICAP report, filled out by any investigative agency participating in the program, "preserve and recall knowledge of similar cases, criminal personality profiles and research studies, preserve information in an active form as a knowledge base, rather than as a mere passive listing of facts and figures, create and preserve a system that is not subject to human failings, will respond to constant streams of data, and can generalize large bodies of knowledge." The hope was that profilers would be able to "receive advice and consultation from the expert system on new and existing cases based upon prior knowledge captured by the system."

Icove hoped that the computer-generated profile would be as accurate as those provided by the Arson Information Management System (AIMS), an FBI crime-pattern-analysis computer program that has enabled investigators to predict the times, dates, and locations of future arson fires. As discussed in Chapter 6, however, the problem of profiling serial killers is far more complex than merely matching patterns; arsonists tend to work in a geographically small area, while serial killers, especially highly mobile, organized killers, often travel over great distances, moving at random intervals.

Although the BSU does not use this system for profiling active cases, Alan Burgess did say that it has proved helpful in training potential profilers by giving them a large number of practice cases and a large knowledge base in a relatively short period of time.

ANYONE CAN CREATE A COMPUTER PROFILE

With the masses of information about individuals available in various computer databases, it is not hard to create statistical profiles of subsets of the population. As early as 1986, the OTA report "Electronic Record Systems and Individual Privacy" stated that at least 16 federal agencies had created computer-generated statistical profiles of the U.S. population. These profiles can be helpful in targeting subsets of the population for more thorough examination for possible fraud, misrepresentation, or abuse of federal programs.

For instance, the IRS has created a profile of categories of taxpayers who are more likely to be underreporting income. Accounts report that one of these categories involves taxpayers who file a Schedule C—profit or loss from an unincorporated business—with a gross, or top-line, Schedule C income of more than $100,000; these people are between three and four times more likely to be audited than are other taxpayers.

As another example, the Social Security Administration has created a profile of those most likely to have errors on their applications for Supplemental Security Assistance (SSA) benefits; characteristics include earned income, home ownership, being age 26 through 40 years, recent separation, and having a bank account.

Profiles can be simple and relatively benign. Many companies' marketing campaigns aim at people who live in certain zip codes, subscribe to certain magazines, are in a certain age group, and have a certain income. These campaigns are often derived from information stored in large databases. There are two problems with this use of databases.

One problem is that the information was often collected for another reason, stored in a database, then sold. Companies are constantly selling their databases to each other, and there are a small number of companies that do nothing but collect varied bits of data and collect them into super databases. These huge databases are used mostly to check credit, but often for other reasons, as well.

The second major problem, possibly larger than the lack of privacy of personal information in these databases, is the lack of security. In an effort to create income streams for prison programs, many states have gone beyond establishing the normal manual-labor types of businesses—furniture, commercial laundry, and so on—that many prisons have run. Specifically, they have set up businesses in prisons to do "back-office" computer-processing chores.

Some telemarketing and tourism-reservation work is now done by prisoners, especially in the south and southwest. Usually, this works out well: Offenders who are low- and sometimes medium-security risks—who all will be released from prison relatively soon and will need to work when they are released if they are to rejoin society successfully—are taught usable skills.

Occasionally, however, this kind of program proves disastrous, such as when sex offenders misuse their computer access to collate data from a number of databases and create databases of likely targets, or even of children whose images, locations, and details of their lives can then be sold to pederasts who frequent the Internet.

A less alarming but still troubling issue is whether those who have assaulted, robbed, burgled, or otherwise injured people should be allowed access to information about unknowing and perhaps naïve callers' travel habits, size of home, purchases, and likes and dislikes. Thus, the use of profiles can also be complex, inherently suspicious, and even damaging.

In addition to being used to look for fraud or misrepresentation within a single program, computer-generated profiles often are used to determine the set of records requested in a computer match. A *match* is a comparison of two or more sets of computerized records in which the matching software is asked to look for patterns or common characteristics. An identifier, usually a Social Security number, is used to match the records—called a *hit*. Hits between the two files must be verified by hand to determine that the individuals really are the same.

Computer matching started at the state level in the early 1970s, when Aid to Families with Dependent Children (AFDC)—the basic state welfare grant—programs began comparing their lists with state labor department wage-earner records; mothers who collect AFDC grants are not allowed to hold jobs outside the house. The federal government got into the act in 1977, when the Department of Health, Education and Welfare (HEW) (now Health and Human Services) announced Project Match, under which HEW compared computer tapes of welfare rolls and federal payrolls in 18 states to detect government employees who were fraudulently collecting AFDC. Project Match, which eventually turned up 7,100 federal employees who were possibly collecting AFDC and not eligible to do so, was decried by many as a "giant fishing expedition."

In December 1982, the Senate Committee on Governmental Affairs' subcommittee on Oversight of Government Management held hearings on computer matching, at which the chairman, Senator William Cohen (R-Maine), who became Secretary of Defense in 1997, reported that "as of January 1982, Federal agencies had completed more than 85 matching programs and state governments are now performing approximately 170 matches involving public assistance records, unemployment compensation records, government employee files, and in some cases, the files of private companies." A 1985 General Accounting Office study put the figures at 126 matches at the federal level and over 1,200 at the state level. The OTA estimated in 1986 that at least 7 billion records had been subject to computer matches just at the federal level.

Matches ran the gamut from federal employees who had also defaulted on student loans to doctors billing Medicare and Medicaid for the same services. A number of matches were carried out in the Department of Agriculture, which administers the Food Stamp program. A disproportionate number of matches were aimed at poor people, who, during the Reagan Administration, were seeing their benefits systematically cut or the criteria for eligibility tightened. As of 1986, no good figures had been generated on the costs versus the benefits of this massive matching exercise.

While statistical profiles may help federal managers to ferret out fraud and abuse of programs—and, consequently, save taxpayers money—there are some dark sides to them. One is that crude profiles or those not well thought out can categorize people on the basis of characteristics that are discriminatory in legally impermissible ways: on the basis of race, religion, gender, national origin, or disability. For instance, one of the characteristics of the

DEA's drug-courier profile is that the person appears to be Hispanic; this can be apparent by surname or even complexion.

A second problem is that merely by characterizing people based on subjective criteria and then singling them out for special treatment, statistical computer-generated profiles violate the equal-protection clauses of the Fifth and Fourteenth Amendments, dictating that all people be treated equally under the law.

Using computer-generated profiles for eligibility programs also violates the due-process clauses of the Fifth and Fourteenth Amendments because these profiles limit the discretion of investigators, caseworkers, and others who are supposed to make determinations, without setting up rules and guidelines for taking away that discretion; those rules and guidelines must be legislated or at least handed down in written administrative rulings.

In some instances, the computer-generated profiles have triggered computer-generated letters notifying claimants that their benefits have been denied or cut off. As Senator Cohen stated in the 1982 hearings, "We have profiles that have been developed by computer, and disability payments that have been discontinued with no human contact coming about until such time as those cases are appealed to an administrative law judge. Two thirds of the cases appealed are being reversed."

What Senator Cohen did not say, because he did not know it until a few years later, was that the Reagan Administration was not using the ruling of one administrative law judge as precedent in any future case, but was forcing every beneficiary denied Social Security disability payments to appeal the ruling. It even continued to deny benefits after those rulings and forced beneficiaries to seek relief in federal court. The administration even went so far as to appeal court verdicts in every federal appeals jurisdiction, using only an appeals-court ruling as a precedent on future disability rulings, and then only in that court's jurisdiction.

Finally, as the OTA put it, "Regardless of their complexity

and formality, profiles by definition are prone to some degree of error, as they are merely probability statements."

Because we use our Social Security number so often—and because we have been so conditioned to think of the democracy in which we live as inherently benign—we may have had (or may in the future have) information about us passed among various state and federal agencies, and we may have been (or may be) categorized as a likely lawbreaker, regardless of our personal ethics. The only way to find out about the uses to which our records have been put in the past is to file requests under the Freedom of Information Act and the Privacy Act with *every agency* and ask whether our records have ever been included in a computer match or a computer front-end-verification program or have ever been pulled as a result of a computer-generated profile; we must then hope that the agency responds truthfully.

The alternative is to admit that we have abrogated many of our responsibilities and handed over many of our rights as citizens to computers—and try not to worry about it.

9

CONTROLLING CRIMINALS: SCIENCE AND CORRECTIONS

Ted Bundy, after years of appealing his sentence in the murders of a Florida State University coed and a 12-year-old girl, was electrocuted in January, 1989. Albert DeSalvo, after admitting to being the Boston Strangler, killer of at least 11 women, was confined to the Bridgewater State Hospital's unit for the criminally insane; a criminal trial was never held.

Bundy was found competent to stand trial and put up a vigorous defense, although his courtroom demeanor often suggested his delusional state: He constantly refused to cooperate with his attorneys, at times acted as his own attorney, had numerous temper tantrums in the courtroom, and sometimes refused to leave his cell for court appearances. DeSalvo, after a lengthy set of hearings, was found incompetent.

How can this be? If the FBI Behavioral Sciences Unit's research is valid, and criminality—even at its most viciously deviant in the form of rape, murder, or mutilation—is more an issue of outlook than of mental illness—shouldn't all serial killers and other violent criminals be treated the same? Also, if the BSU's research is valid, is there any way to change a criminal, to help him or her to create a new outlook?

Throughout U.S. history, public opinion has vacillated about how criminals should be treated. At different times, social policy has emphasized punishment, rehabilitation, or incarceration. Various aspects of science and technology have been used to carry out political and social mandates in each of these areas.

The technology of punishment in the seventeenth and eighteenth centuries, such items as the stocks and the dunking chair—resembled medieval instruments of torture. In the eighteenth century, the Quakers of Philadelphia created the first penitentiary, the Walnut Street Jail, to which convicts were sentenced to spend time rather than undergoing corporal or capital punishment. The penitentiary system signaled the change from an era of punishment to one of rehabilitation. The Quakers were encouraged in this undertaking by a 1778 English description of the purpose of a penitentiary, which called for "sobriety, cleanliness, and medical assistance," a "regular series of labour," solitary confinement, and religious instruction as a way to restore offenders to good habits and rid them of "pernicious company."

By the mid-nineteenth century, training schools were a popular vehicle to take delinquent and criminal youngsters off the streets of burgeoning cities and teach them trades they could presumably use. In the twentieth century, U.S. prisons have been alternately warehouses to incarcerate criminals and laboratories for sociologists and psychologists to try to find ways to get criminals to understand their criminality and develop ways to cope when they reemerge into society. In addition, as psychiatry came into its own throughout the century, and psychiatric diagnosis and treatment became more precise, the question of competency and ability to treat the criminally ill was added to the mix.

In much the same way that scientists argue over whether criminals are born or made and whether genetic, environmental, or larger societal factors cause crime, the argument over prisons often revolves around whether they deter crime, help criminals sharpen their criminal techniques by putting them in contact with criminals with other skills, or merely keep criminals off the streets for varying periods of time during the most crime-prone years of their lives—typically between about ages 15 and 35 years.

In the 1980s and into the 1990s, while many people were advocating the building of enough prisons to lock up anyone convicted of even modest crimes, and others were continuing to argue for the "liberal" alternatives to prison such as restitution and community service, still others were looking to science and technology to solve the problems of prison crowding.

Bird Dogs and Electronic Leashes

In the same way that an electronic location monitor can be put on a person or vehicle under surveillance, an individual convicted of a crime can be monitored constantly by use of an electronic "bird dog."

The first item of this kind, created by Ralph Schwitzgebel, a Harvard University behavioral scientist, was patented for use in 1969. Simply, this form of electronic parole was a small transponder the convict carried, which relayed information about his or her whereabouts to a transceiver located in the parole office.

Immediately, civil libertarians decried the possibilities of being able to monitor a person's actions; a lot can be learned about a person by knowing where he or she is at any given time, even when the person is engaged in innocuous, legitimate activity. On the other side, law-and-order advocates called for adding to the simple electronic monitor some additional technology that would transmit anything the offender heard or said, readings taken regularly from sensors that recorded physiological data such as heart and pulse rates and blood-alcohol levels, and other forms of minute-by-minute surveillance. Fortunately, this kind of anthropotelemetry was not carried out in any large-scale way.

In fact, the idea of electronic parole fizzled for more than a decade, until prison crowding began to be a serious concern in the 1980s. Since then, dozens of local and state electronic-monitoring programs have been put in place.

One of the first electronic-monitoring programs was instituted by Judge Jack Love in the Second Judicial District of New Mexico in 1983. Love allowed those convicted of drunk driving and some white-collar crimes to spend time on electronically monitored probation, rather than in jail; the program was upheld by the state supreme court, as long as the "privacy and dignity" of proba-tioners and their families were maintained.

Within a couple of years, correctional agencies and private probation-monitoring corporations were running programs in Florida, Idaho, Kentucky, Michigan, New Jersey, Oklahoma, Oregon, Pennsylvania, Texas, and Utah. By 1989, more than half the states in the country had at least one pilot program in place.

More recently still, some jurisdictions have turned to elec-tronic monitoring of people on bail, awaiting trail. On July 27, 1989, Saudi financier Adnan Khashoggi was released in New York on bail—and wearing a monitoring bracelet—to await trial on charges of helping former Philippine President Ferdinand Marcos to move millions of dollars out of the Philippines into overseas investments.

In these programs, individuals may not be under 24-hour supervision, but they must spend most of their time under house arrest and often are allowed to leave the house only to work or go to school, to attend meetings with a rehabilitative purpose such as Alcoholics Anonymous, and to perform personal chores such as shopping; usually, the person must perform out-of-the-house chores during preset times.

These electronic house-arrest programs are of two types:

active or passive. In a so-called *passive program,* the arrest is enforced by telephone calls to the individual, made by either a person or a computer. In a computerized system, the call must be answered both by voice and by electronic identification. The person must both answer the phone and insert a wristband electronic encoder into a verifier box attached to the telephone. If either the phone is not answered or the connection is not verified with the encoder, a violation is recorded by the computer monitoring the calls.

In an *active program,* the person wears either an ankle- or a wrist-bracelet transmitter about the size of a cigarette package, which usually allows him or her to move only dozens or hundreds of feet from home. This transmitter sends an encoded signal at regular intervals to a receiver inside the home, which retransmits the signal to a central computer. The police station or parole office has a communications system that receives messages from a number of central computers to alert them to any violations.

Electronic parole is more expensive than conventional parole, but far less expensive than incarceration. Because parole officers' caseloads are as overburdened as prisons—overburdened parole officers have helped keep the cost of traditional parole programs below the cost of electronic-monitoring technology—electronic parole provides far more supervision than does traditional parole, although it lacks human contact and lacks the parole officer's ability to sense subtle changes in an individual's behavior that could signal a turn for the worse.

The overriding question in electronic-parole systems is, For whom is it appropriate? Most people would not put a violent offender or repeat serious offender on electronic monitoring, fearing that he would slip his leash long enough to commit other crimes; although, as newspaper articles constantly remind us, people on traditional parole also often commit repeat crimes, as do those who participate in prison furlough programs.

Beginning in 1994, New York City turned to technology to perform "triage" on its probation system. Before that time, all of the 60,000 or so individuals on probation were required to make brief (5 to 15 minutes) face-to-face appearances with a probation officer who had a caseload of anywhere from 500 to 800 individuals. Some met every week, some as infrequently as twice a year.

Under a new system, only the most violence-prone individuals would meet with their probation officers regularly, sometimes for as much as four hours a week in hourlong daily sessions, while the vast majority would merely report to an automated probation kiosk that looks like an automated teller machine. Checking in weekly at the machine is intended to make sure the probationers remain around the city. The machines contain information about the probationer's housing and employment status, which can be verified or updated. The city tested voice-recognition technology and recorded some users, in order to make sure it was actually the probationer doing the reporting.

Michael Jacobson, the city's probation commissioner at the time, said the effort had three goals:

1. To better utilize scarce probation officer resources, focusing on those who needed more close supervision. (The aim is to keep a tight leash on those who have shown violence in the past and keep them from committing more violent crimes and landing in already overcrowded prisons.)

2. To reduce costs, in light of signs from the state government that funds for probation and parole would be cut over both the short and the long term.

3. To engender a different attitude, by both probationers and probation officers, about the effectiveness of probation as an intervention.

By limiting eligibility to those convicted of misdemeanors and minor, nonviolent felony offenses, however, U.S. taxpayers may be paying more than we need to because many of these people would play by the rules of traditional parole (which is cheaper). The worst danger is that we will delude ourselves into thinking that because this technology is simple, effective, and rather inexpensive, it should be used to punish any behavior that is mildly socially aberrant or offensive; we could be allowing what is in effect minimum-security technology to help turn the country into a maximum-security society.

Biochemical Corrections

Despite numerous objections to biochemical and nutritional treatment to modify behavior, it is done in some instances, most notably with a hormonal program for sex offenders. The numbers of recorded rapes rose steadily through the 1970s and 1980s, then leveled off in the 1990s. Murders have actually declined slightly in the 1990s from the peak of almost 25,000 annually in the late 1980s, at the height of drug wars in many cities. Rape statistics are thought to reflect a better understanding of the problem and a greater willingness of victims to come forward and of police to classify questionable sexual activity as rape. Regardless, the amount of nonconsensual sex that takes place in this country is disturbing, especially when it relates to sex with children.

The criminal-justice system has always had a difficult time distinguishing between the sex offender who is evil and the sex offender who is sick and determining how to deal with both. Among the varying types of sex offenders are those who are violent, often driven by rage and anger, and who use sex as a weapon of power. Other offenders deny the criminal nature of their sexual activities; still others admit their criminality, but say they were disinhibited by alcohol, other drugs, stress, or other factors.

Still others, classified as paraphiliac, are sexually aroused by either the fantasy or the actual carrying out of a particular deviant sexual activity; this activity may be as mild as transvestitism or voyeurism, or as violent as rape. Many paraphiliacs, nearly all of whom are men, lack sexual-impulse control. Many paraphiliacs are adults with families; some conduct their deviant behavior outside the family and are able to hide it for a long time. Over one third of all convictions for sex crimes are for exhibitionism, and another third or more are for sexual activity with children (the desire to have sex with children is clinically known as *pedophilia*), leading one to think that there are far more sick sex offenders than there are truly evil ones.

Once the criminal-justice system has determined a particular person to be deviant and in need of treatment, rather than evil and in need of punishment, the question is how to treat that person. Most programs combine psychotherapy, life-skills training, and behavior modification. The most controversial part of some of these treatment programs is the use of a hormonal compound called Depo-Provera, a synthetic progesterone similar to the chemicals cyproterone and cyproterone acetate, which have been used to treat male sex offenders in Europe since the 1960s. Both the natural and synthetic versions are *antiandrogenic,* meaning that they suppress the production of testosterone, a hormone found in both men and women, which is responsible for the male sex drive. According to the Office of Technology Assessment, 14 percent of adult sex-offender treatment programs and 6 percent of programs for juvenile offenders use Depo-Provera.

The use of Depo-Provera or other antiandrogenic progesterone treatment often is called "chemical castration." The goal in antiandrogen treatment is to lower the level of testosterone from that found normally in men (between 400 and 1,000 milligrams) to that found normally in women (between 40 and 100 milligrams). The outcome hoped for is a reduction in the frequency of erotic fantasies and diminished sexual desire. The antiandrogenic progesterones are used because they have far fewer unpleasant side effects than do large doses of the main female sex hormone, estrogen—which, when administered to men, causes body changes such as reduction in body hair, the enlarging of breasts, and possibly irreversible infertility. Depo-Provera does have some side effects, including fatigue, depression, weight gain, and headaches.

Patients in the United States being treated with Depo-Provera usually receive injections of 300 to 400 milligrams about once a week, depending on their weight and general size. Researchers who have used the treatment report that when it is used in conjunction with traditional psychotherapy and behavior therapy, there are almost immediate behavior changes, noticeable to the patient, and far fewer of these patients become discouraged by lack of progress in treatment than do those who undergo treatment without the hormonal component. Many scientists report that although Depo-Provera has no effect on violence and aggression in general, they do see a significant reduction of sexual aggression among patients, suggesting that the treatment works well with paraphiliacs, but not with antisocial criminals for whom sexual crime is but one form of criminal outlet.

Researchers report that for Depo-Provera therapy to be successful, it must be carried out over the long term; within a year or so after the last dose is administered, men's testosterone levels return to normal. Although programs have shown little recidivism during the treatment period, the period following short-term use shows a high level of relapse into sexually deviant behavior.

As with other methods of alternative sentencing, Depo-Provera treatment raises significant questions for both those who are treated and those who are not.

Some see this "chemical castration" as cruel and unusual punishment, not unlike physical castration, lobotomy, forced sterilization, or other physiological treatment methods previously used in this country to try to correct deviant behavior, whether administered in a punishment or a treatment setting. Others argue that because entering a Depo-Provera treatment program is usually a condition of parole, it is coercive, given that sex offenders—especially those whose predilection is sex with children—often are harshly victimized in prison. Some practitioners view this use of Depo-Provera as a violation of professional ethics; when seen as biochemical experimentation on a population that is not free to refuse—even though a person may withdraw from the program, the knowledge of what awaits in prison can be psychologically overpowering. Also, these parolees may not be subject to the rules of informed consent accepted in medical experimentation; there is some evidence that long-term use of Depo-Provera could be carcinogenic. Visions of such grotesque experimentation as the nontreatment of black prisoners with syphilis in some prisons up until the 1950s, or even the experimentation carried out during World War II by Nazi doctors, alarm many people.

On the other side of the coin are those who argue that Depo-Provera treatment is unconstitutional because it is not equally available to all prisoners as a substitute for prison or as necessary medical treatment, and those who are best able to get into programs are articulate, middle class, and white. The same kinds of objections often are raised in arguments about the merits of another drug therapy, disulfiram, known as Antabuse, which is used in the treatment of alcoholism.

Although alcoholism clearly is regarded as a disease, its manifestations often are criminal, especially in the form of drunken driving and the dangers to others that this causes. In addition to the traditional treatments of alcoholism such as counseling and self-help support such as Alcoholics Anonymous, many treatment programs, including those that are alternatives to prison for those convicted of alcohol-related crimes, have turned to Antabuse.

If an individual takes an Antabuse tablet, he or she must not drink alcohol for at least 72 hours; the result of mixing Antabuse and alcohol is violent nausea and vomiting, drop in blood pressure, blurry vision, and even difficulty breathing. This chemical aversion softens up the drinker and makes him or her more responsive to the other therapies. It takes only one reaction to put fear into the drinker and, again, many opponents argue that the fear of the reaction, combined with the fear that the often middle-class, articulate offender has of going to prison, forms a highly coercive punishment disguised as treatment.

As other forms of behavior modification proliferate in experimental settings, no doubt the same objections will be raised. How the courts will respond is difficult to tell; so far, the federal courts have seen cruel and unusual punishment under the Eighth Amendment as torture or lingering death, punishment that is "disproportionately severe," extremely poor living conditions in confinement, and even the use of some forms of aversion stimuli in people involuntarily committed to mental-health facilities. The courts have ruled that prisoners must consent to their treatment and must be given the option of leaving the program. Much the same rights have been given to those committed to mental-health facilities, although some professionals argue that because individuals confined there are such a danger to themselves or those around them and because they are unable to make any rational judgment, there is room for nonconsensual treatment.

CHANGING THE CRIMINAL'S OUTLOOK

Beginning in 1961, when he took a position at St. Elizabeth's Hospital in Washington, D.C., a federal mental-health facility with a large population of the criminally ill, Samuel Yochelson, a psychiatrist, developed a program to help criminals change their outlook. That program was taken over in the late 1970s by Stanton Samenow, author of *Inside the Criminal Mind* and, since the late 1980s, a psychologist in private practice in Virginia.

Samenow describes the program through a story about Leroy, a criminal confined to the hospital:

> Having heard that Yochelson's group met all morning every day, Leroy wondered what in the world consumed

so much time. He quickly found out. First, Leroy had to be taught to stop and recollect what he had thought, then to make notes on paper. He was instructed to think of this exercise as though a tape recording of his thinking were being played back. The reason for this emphasis on thinking was that today's thoughts contain the seed of tomorrow's crime.

Yochelson and Samenow argue that criminals see themselves as victims, victims of all the picayune rules of behavior that the rest of us, for the most part, feel comfortable living within. Criminals also live in a state of perpetual anger at those rules of behavior, as well as at the people who seem to function and even prosper within those rules.

As argued in the FBI BSU's report on "The Men Who Kill"— a series of interviews with 36 serial murderers, assassins, and other brutal killers—"Criminals are totally self-centered" and lack the ability to see the world in any terms except how they are hindered. Because of this, they feel it is their right to take whatever they want—or whomever they want—whenever they want it. The criminal does little, if any, planning and if anything goes wrong, others are always to blame.

Yochelson, and later Samenow, worked with small groups of criminals over a period of months, or even years, to get them to talk about their feelings and learn how to take control of their feelings and the situation, to set goals, and to work through temporary obstacles. As with any other situation that one can force oneself to overcome, there are stages of learning and action. Samenow calls this method "the program," and in its full description, it resembles other self-help programs such as Alcoholics Anonymous.

In his book, Samenow argues that long-term adherence to "the program" of critical self-evaluation can change a criminal's outlook. Unfortunately, the level of intensive resources needed to provide such a program for all the criminals who need it would overwhelm the country's resources—although hiring enough police to catch criminals; hiring enough judges, prosecutors, and public defenders to make the court system run more efficiently; and building enough prisons to house convicted criminals also may bankrupt us.

On the one hand, we must break through our societal mistrust of medical and psychiatric expertise, the notion that psychiatrists and psychologists see a diagnosis where there is only a poor attitude; on the other hand, we must understand that helping someone to cope better with societal norms is not necessarily taking away that person's autonomy.

Postsentence Confinement and Community Notification

During the 1990s, two other correctional practices for sexual offenders have become steeped in controversy. One is continued confinement in a mental-health institution after a prison sentence is completed. The other is community notification regarding sexual offenders. Both practices have come about as the result of a few highly publicized incidents where sexual offenders, released after serving a sentence, have victimized children. Both practices have been ruled constitutional by federal courts.

Community-notification laws began in 1994, when the state of New Jersey passed "Megan's Law," named after 7-year-old Megan Kanka of Hamilton Township, who was raped and murdered by a neighbor, Jesse Timmendequas, who had recently been released after serving a prison sentence for child rape. The notification law the case spawned was passed before Timmendequas was even tried. He was eventually convicted.

It is ironic that Timmendequas was the roommate of Joseph Cifelli, another child molester, who grew up in the house that he and Timmendequas were living in at the time of Megan Kanka's murder. The two had met at Avenal Prison, the site of New Jersey's sexual-offender treatment program. While few neighbors knew Timmendequas, most knew Cifelli and knew his history.

Megan's Law, and similar laws passed in more than half the 50 states by 1998, mandate that to one degree or another, communities be notified when a convicted sex offender moves into the community. This notification is made after the offender registers with the local police where he (sex offenders are predominantly men) will live. Many times, a condition of release or parole is for the offender to live in a particular place, often with family.

The amount of notification and the means of notification are different in each state. Some states require that notices be placed in local newspapers. Some mandate that public displays of the photos and addresses of sex offenders be posted at civic events such as carnivals and fairs. Others require that police provide personal notification by letter, sent to immediate neighbors and to all those in the neighborhood with children.

Suits have been filed against the laws, claiming they are cruel and unusual, in that they leave the offender open to harassment and possibly violent assault at the hands of mobs. Individuals and groups that have held vigils outside the homes of released sex offenders have been sued by the offenders for harassment, and by their estates after offenders have committed suicide, possibly out of fear or psychological torment.

Most notification laws were enacted in 1994 and 1995 but did not take effect for one to two years because of suits pending in each state. By the end of 1997, federal courts had ruled these laws constitutional or had sent them back to state legislatures for modification, then ruled the modified laws acceptable to constitutional standards. A few cases had been taken to the Court of Appeals, but by the end of 1997, it was clear that community notification regarding sex offenders, when done within certain parameters, is a constitutional form of postpunishment monitoring. Sex offenders who are harassed, assaulted, or driven out of the community will have to seek redress through civil court. Very few have succeeded.

In June, 1997, the United States Supreme Court ruled a Kansas law constitutional that confines repeat sex offenders to mental-health facilities after their prison sentences are up. The case that went to the Supreme Court involved Leroy Hendricks, a 62-year-old admitted pedophile, who once told a court at trial that the only way he would stop assaulting children was to die. The Kansas law and others that sprang up in the 1994–1995 time period allow states to transfer "dangerous" sex offenders to mental-health facilities at the end of their prison sentences if their offenses are committed because of a "mental abnormality." The state must hold a hearing on the matter; at this hearing, the state has the burden to prove that the offender it wishes to confine is a "sexually violent predator."

Hendricks argued that his confinement after finishing a prison sentence would subject him unconstitutionally to double jeopardy—two punishments for the same crime—and ex post facto law—a new punishment for a previous crime. Both, his attorneys argued, were violations of his due-process rights under the Fourteenth Amendment.

The Kansas Supreme Court held the law unconstitutional, but the state appealed directly to the U.S. Supreme Court, which ruled 5:4 that due process was not violated. The court ruled that the postsentence confinement was not punishment, but rather a novel approach by the state to use civil involuntary commitment to a mental-health facility. Hendricks's attorneys had argued that because their client was not receiving treatment, the confinement constituted punishment.

At the time of the Supreme Court decision, only five other states—Arizona, California, Minnesota, Washington and

Wisconsin—had similar laws on the books. Since the Supreme Court ruling, a few other states enacted such laws, and many states anticipated attempts to enact such laws in 1998 and 1999.

Eric Janus, a professor at William Mitchell College of Law in St. Paul, Minnesota, who led the challenge against the Minnesota law, argued that the new civil-commitment laws are reminiscent of the "sexual psychopath" laws of the early twentieth century, which were applied to people who were seen as deviant, such as exhibitionists and consenting homosexuals.

Kay Jackson, a New York psychologist who treats paroled sex offenders, told *The New York Times* in response to the Supreme Court decision, "We have not decided to incarcerate them for life or kill them. We have a big social dilemma about what to do with these people."

10

LIBERTY, JUSTICE, AND SCIENCE

The U.S. criminal-justice system is a mass of conflicts and contradictions. While trying to protect innocent people from criminals, we also must try to protect the innocent from being mistaken for criminals. Science should be purely impartial in this process—it should prove as much as possible that pieces of physical evidence are what they are supposed to be and that they came from where witnesses and investigators say they came from. Unfortunately, in the U.S. legal system's adversarial approach, there really is no such thing as impartiality of evidence.

As science becomes more complex, the need for experts to translate the meaning of science becomes more acute, but not even these experts are impartial: Both the prosecution and the defense in criminal trials, and both parties in civil trials, routinely have their own scientific experts testify on nearly every aspect of the evidence. This leads to the question of whether it is possible to find an expert who will defend any position—as an attorney does—or whether there is so much scientific uncertainty in physical evidence that it really is not possible to use any such evidence in a conclusive manner.

Doctors, lawyers, and scientists are becoming increasingly involved in discussions of how science can best be dealt with in the courtroom. The discussion began in earnest in February, 1989, when two law-school professors and a professor of public health held a seminar about improving procedures for examining scientific evidence in *toxic torts,* lawsuits claiming health or property damage from allegedly toxic substances. Examples of such lawsuits are those surrounding Agent Orange, asbestos, silicone breast implants, and even cigarettes. The main problem then was that there is no systematic way to evaluate evidence provided by experts; whichever side is defending the position that runs counter to generally accepted scientific evidence usually can find an expert to exploit gaps in that body of evidence. Judges and juries find it hard to argue against novel scientific theory; who's to say, after all, that an unpopular scientific view—as the notion that the earth revolves around the sun was a millennium ago—isn't correct.

E. Donald Elliot, a professor at Yale Law School and one of the seminar organizers, told *Science* magazine that the adversary process "extends equal dignity to the opinions of charlatans and Nobel Prize winners."

Marcia Angell, managing editor of the *New England Journal of Medicine,* wrote her 1996 book, *Science on Trial,* focusing on just this issue, as it pertained to the class-action suit against the makers of silicone breast implants. In that case, a multibillion-dollar class-action settlement was agreed upon during a trial in which much anecdotal evidence was presented that silicone breast implants "cause" connective-tissue disease. Much of this evidence was presented by self-proclaimed medical "experts" who attributed the disease of an individual woman to an implant.

Over the years during which the suit was in process, however, and in the few years after the settlement, a number of scientific studies found no statistical evidence that the incidence of connective-tissue disease was any greater for women who had had silicone breast implants than for those who had not. What happened in the breast-implant suit, and what is continuing to happen in tobacco litigation, has caused judges, especially on the federal court, to increasingly appoint special scientific masters to hear scientific evidence away from the jury, and to rule on which scientific evidence has enough credibility to be heard by the jury. Although federal judges have always had this power implicitly, the practice was challenged; the U.S. Supreme Court ruled in 1997 that it is not only valid, but also desirable; they urged federal judges to use the procedure more often.

The scientific experts called to testify in 1989 in the Castro case in the Bronx—who held their own miniseminar on both the science involved in the case and their role as expert witnesses—had two basic arguments against the reliability of the methodology used by a private laboratory in conducting forensic DNA typing. One was the lab's contentions about population statistics and scientific uncertainty. The other was sloppy lab work—a charge that also has been leveled frequently at public forensic labs across the country, even by the Department of Justice.

It is clear that some people are in prison today because of bad science done in the name of criminal justice: bad laboratory procedures; exculpatory evidence withheld from a defendant because of continual public and political pressure to gain convictions; and scientific evidence that is far from certain, but which has been argued persuasively by prosecution experts.

More shocking than these careless or overzealous actions by members of the criminal-justice system are instances of willful falsification of evidence. On September 21, 1992, Ralph Erdmann, who worked for years as a freelance medical examiner in rural Texas, pleaded no contest to seven felony counts of falsifying autopsies in Lubbock, Hockley, and Dickens counties. The 65-year-old physician was sentenced to 10 years probation and 200 hours of community service, and he lost his license to practice medicine.

Erdmann was accused of faking up to 100 autopsies, botching dozens of blood tests, and filing false paperwork on countless other autopsies, including false toxicology reports. Attorneys in the three counties where Erdmann worked as a contract medical examiner, performing as many as 400 autopsies a year and earning over $200,000 some years, appealed dozens of cases, including 20 capital cases, after the charges were brought. Texas courts in 1997 were still sorting out new trials and appeals of those convicted based on Erdmann's evidence. Although some defense attorneys argued that prosecutors in the three counties had known of Erdmann's malfeasance, no charges were ever brought against them.

In 1993, a West Virginia court said that at least 134 prisoners in that state might be entitled to new trials because of invalid serology tests carried out by the state police chief serologist, Fred Zain. Zain was the state's top expert on blood and semen stains from 1979 to 1989. The court ordered state police and prosecutors to continue holding evidence from 134 rape or murder cases between 1986 and 1989 that Zain was involved in, in case new trials were ordered.

Zain first came under suspicion in court in 1987, when he was accused of conducting flawed tests showing that Glendale Woodall was 1 of only 6 in 10,000 people who could have committed a rape, for which he was convicted and sentenced to 365 years. Later DNA tests showed conclusively that Woodall was not the assailant, and he was freed in 1992. After Woodall's release in 1992, West Virginia's Supreme Court of Appeals appointed a special judge and panel of experts to examine Zain's work. During that investigation, they found that two state police officers who worked under Zain had complained about his techniques as early as 1985, but that nothing had been done about it.

After leaving West Virginia in 1989, Zain was hired in Texas as a serologist in the Bexar County Medical Examiner's office. He was later dismissed from that job after complaints about his doctoring of results. Zain was later convicted of filing false reports.

In yet another case of forensic overzealousness, Claus P. Speth, a highly regarded forensic pathologist, was convicted of manufacturing evidence while working as a private pathologist. The case involved the death of 43-year-old Ronald Puttorak, found hanging in a cell in the Essex County, New Jersey, jail in August 1993. The state medical examiner's office—where Speth had worked in the 1980s—ruled the death suicide by hanging. Speth, working for the family in a federal suit against the county for violating his civil rights by not supervising the jail and allowing him to be killed, argued that Puttorak had been strangled in his cell.

The case turned on the tiny hyoid bone, a U-shaped bone in the front of the neck, which supports the tongue.

State pathologists said the hyoid bone was intact, consistent with hanging by a rope with a knot in the back; but Speth said it was fractured and there was trauma around it, consistent with strangulation by hand. A jury found Speth guilty of evidence tampering.

Finally, just before the federal government began presenting evidence from explosives experts from the FBI laboratory in the 1997 case of Timothy McVeigh, the Justice Department Inspector General's office issued a blistering 500-page report on the inner workings of the various portions of the FBI lab where bomb analysis is done. The report highlighted political infighting, as well as sloppy forensic work by some personnel. Although the report did not keep the government from convicting McVeigh, or his coconspirator Terry Nichols in a later trial, it did cause much anxiety not only for the prosecutors in the Oklahoma City bombing cases, but also for those who later prosecuted the Unabomber in 1998. The then-ongoing FBI investigation of the explosion of TWA Flight 800 in 1996 was also thrown into shadow by the report, and the FBI Director Louis Freeh ordered a major shakeup of the lab, including reorganization of which groups work on bomb-analysis investigations and the recruitment of a new lab director from outside the FBI. We can only hope that diligent judges will continue to force juries to take into consideration more than just circumstantial scientific evidence when making a conviction and insist that there be motive, opportunity, and, it is hoped, witnesses.

Regardless of how good or bad the scientific evidence is in any individual court case, the use of ever more sophisticated scientific techniques in the name of justice raises larger and more troubling questions:

- If there is a scientific method for extracting information about a person from his or her body tissue or fluids, when is it legitimate to use that method, and when does using that method constitute a violation of Fourth Amendment rights to be free from unwarranted search and seizure, and Fifth Amendment rights not to incriminate oneself?

- How much and what kind of information should be kept about U.S. citizens, both those convicted of criminal offenses and those never convicted?

- Who should have access to that information? For what purpose?

- Will there ever be a clear scientific definition of the difference between sickness and evil?

Let's look at a possible scenario. A brutal rape and murder has occurred, and police are searching for suspects. There are no fingerprints and no bite marks on the victim. There are blood and semen stains. By using a number of scientific techniques, including DNA typing, some of the bloodstains are shown to be from the victim. Other bloodstains match the semen stains and can be assumed to be from the killer. There were no witnesses to the crime and no apparent motive.

Where do police start, in order to widen the investigation? One avenue might be to find those who have been convicted of similar crimes in the past. Many, obviously, are in prison or mental hospitals, but some have finished their sentences or have been released from treatment. Others, for various reasons, have served a probation sentence or have been treated in an outpatient setting. This information can be found fairly quickly by checking the arrest, conviction, and sentencing information relating to sex offenses in the state criminal-justice databases.

Each of these people known to be free when the killing occurred, and living within some arbitrary geographic area—the city, the county, or whatever—could be contacted, interviewed, and investigated. Presumably, there would be many with very good alibis, but the number of possible suspects still could be large.

It would be helpful at this point to have each of these men give blood samples and to have those samples tested to see whether the DNA matches that of the blood and semen samples that presumably are the killer's. Would that tactic be legal? It might.

In some states it isn't even necessary to obtain samples, because every convicted sex offender and murderer gives blood and saliva samples upon entering prison, which are DNA typed. That information is stored in the state's computerized criminal-justice records. Before long, DNA information will be part of the FBI's NCIC records; the FBI will then centrally hold DNA information for those states for which it currently holds other criminal information, and, for those states for which it does not hold information, it will direct queries to the proper record system.

What about a state without preprison deposit and record-keeping? Would the fact of prior conviction constitute probable cause for police to get court orders to do blood tests on perhaps dozens or even hundreds of men? Isn't there a notion in this country that the police should have determined through their investigation that there is some tangible connection between a person and an incident before they can get a warrant to search a person's possessions—and, by extension, his or her fingerprints or body fluids?

Well, the Supreme Court ruled in 1989 that a particular pattern of behavior that fit a profile was enough justification for Drug Enforcement Administration personnel to detain an airline passenger, even though the man had no record of drug convictions. In the Supreme Court's majority decision, Chief Justice Rehnquist wrote that the "totality of the circumstances" allowed the DEA to detain a person on arrival in Hawaii from Florida, solely because he met the criteria of the DEA's drug-courier profile. Rehnquist wrote that the DEA did not need "probable cause," but only "reasonable suspicion," to make the stop and to arrest the man after finding drugs.

The court's decision was an early chip taken out of the doctrine of probable cause, which requires police to develop evidence linking a particular person to a particular incident before they can go before a judge to seek search or arrest warrants. Further Supreme Court rulings since that time have strengthened the government's ability to use predetermined profiles.

Following the Rehnquist logic, prior conviction for a sex offense certainly would constitute a person fitting the profile of a sex offender, and police probably could compel convicted sex offenders to give body-fluid samples under the new doctrine of reasonable suspicion.

It may bode ill for all of us to have a Supreme Court ruling that allows stops, searches, and arrests based merely on reasonable suspicion because a person meets the (by definition) somewhat subjective criteria of a physical profile. Demographers, sociologists, and even advertisers are able to characterize us by our physical and behavioral qualities. The ability to create such profiles—and for the police to act on them—makes it possible to conduct mass roundups of certain kinds of people, based on very limited evidence of any wrongdoing.

If the police were allowed to take blood samples of all known sex offenders without any ironclad alibi for the time of the crime being investigated, what would happen to the information once it was gathered? Would the results of DNA typing in a particular incident be destroyed, or would a permanent record be created—in effect, creating a system of archived DNA typing, only at a greater cost to the public? If the information were destroyed, what would happen the next time there is a rape or other sex crime, where there is a crime-scene sample that has discernible DNA? Do police round up the usual suspects and do the tests again? How many times does a convicted sex criminal have to be rounded up, interrogated, investigated, have his blood drawn, and be found scientifically innocent before he either sues for harassment (could he ever win?) or has a mental breakdown?

Maybe the permanent record really is the better solution, like the permanent fingerprint record. A sex offender should know that he might be able to obliterate his fingerprint—by wearing gloves—but leaving his DNA behind probably is another matter.

Another question: If having a permanent DNA-type record for sex offenders and other violent criminals is a good idea, why not have a DNA-type record for every U.S. citizen? The test costs less than $100. Why not draw blood from every newborn and create a national database of DNA samples—every baby with its own personal bar code? In fact, companies are offering the service to pediatricians and parents as a tool for finding missing children, or for identifying their remains.

Couldn't the argument be made that knowing every person's DNA type would be more broadly useful than just for criminal justice, that there are a lot of times when positive identification is impossible, and that DNA would be a wonderful identifier? Indeed, that argument could be made and has been made. Researchers now are looking at ways to extract enough DNA from bone to do a conclusive DNA typing; if this becomes possible, scientific sleuths could clear up the cases of the approximately 5,000 sets of skeletal remains found each year in this country.

The military already is looking to DNA typing and recordkeeping for its grisly task of positive identification of those killed in action or in accidents. There may never be an "unknown soldier" from any future war.

Nonetheless, a national database of DNA information certainly would be a trade-off. On the one hand, crime might be lessened, criminal convictions might be made easier, and positive identification would be enhanced, but on the other hand, such a "universal identifier" would create a civil-liberties jungle.

Despite notification on numerous documents that the Social Security number is not to be used for identification purposes, it is used in such a way extensively—for instance, on driver's licenses, school records, and medical records. As we've seen in previous chapters, this identifier can be used to compare and match computerized records by governmental agencies in order to create lists of people who fit a number of criteria—the creation of computer profiles. Any third-level bureaucrat, with access to a personal computer and networking equipment, can call up information about a person from dozens of federal agencies just by flagging a Social Security number—such confidential information as tax returns, loan applications, and participation in government grants and social-welfare programs. Beyond government, these records are used by insurance companies, credit-checking agencies, and even advertising and marketing companies.

Imagine the invasions of privacy and the detriments to free-speech rights possible from being able to identify people through DNA, especially if the technology is sensitive enough to type minute samples of hair, saliva, and even finger oils. People handing out leaflets for causes or political candidates, or otherwise engaged in innocuous and perfectly legal activities, could be identified and harassed.

Other methods of tracking and identifying people, such as coded dental implants or biometric sensors, pose the same frightening questions.

Let's broaden the DNA question for a moment. The DNA-typing process was an outgrowth of medical research into finding genetic markers on DNA, alerting doctors that people carry the predisposition to certain diseases. Each year, nearly half a million U.S. workers contract disabling occupational illnesses. Although particular environmental conditions may (or may not) trigger the onset of such an illness, often a person's genetic makeup predisposes the individual to the disease; geneticists call this *hypersusceptibility.*

Could a universal DNA record include not only a DNA fingerprint of an individual for identification purposes, but also information about hypersusceptibilities? Of course. Could that record then be used by industry—and government and insurance companies—to screen potential employees, soldiers, public officials, and policyholders? Theoretically, yes. As we've seen, problems of data control and data integrity are rampant in large databases.

Could the argument be made by government, industry, and insurance companies that such a screening and denial of work or benefits are perfectly logical and legal, indeed ethical? It sure could: If a person knows that he or she would face increased risk of occupational illness and disabling disease by working in a particular industry, shouldn't that person know about it and be counseled to work somewhere else? If that person then refuses to work somewhere else, works in an industry that he or she knows is "hyperdangerous" to himself or herself, and contracts such an illness or disease, whose fault and responsibility is it? Why should consumers bear the burden of higher costs for industry to pay for insurance and treatment? Why should other policyholders pay for insurance for these people? Why should taxpayers pay for treatment for these people if they are poor?

Could the use of a universal DNA identification system create a society where none of us is willing to share the normal risks of living, where only the genetically strong survive? Instead, would we, as a society, seek some "final solution" to rid ourselves of the genetically defective and the hypersusceptible, as well as the sex offender?

Well, we've rounded up our community's sex offenders, tested their blood, and, indeed, found a person whose DNA fingerprint matches that of the crime-scene samples. The evidence has been presented at trial, and the jury believes both the laboratory science and the population statistics, which say that the possibility that this DNA fingerprint belongs to someone else is 1 in 1 billion.

The defense attorney, feeling overwhelmed, turns to one last avenue to keep his client out of prison—the claim that this man is not evil, but sick. To make this argument, the defense probably would depend on an expert forensic psychiatrist who comes from the philosophical position that people who commit bizarre and brutal crimes are, by definition, not normal or sane.

The issue of insanity is an area where science—in this case, medicine—and the law come into conflict. The legal definition of *sanity* is a companion to other legal definitions of *competence*: the ability to reason, to understand the consequences of one's actions, to make decisions for oneself, and to assist an attorney in the defense of one's position. These legal boundaries allow for such decisions and controversies as

- Courts granting severely ill people's requests that they be removed from life-support mechanisms

- A court allowing a severely handicapped man—a quadriplegic—to be removed from life support. (Advocates for the handicapped had argued that this set a dangerous precedent because the court was acceding to the wishes of a depressed, but otherwise healthy man who could live for years on life support, as opposed to someone for whom life support was, in effect, postponing death.)

- Allowing Theodore Bundy not only to stand trial in the brutal murder spree on the University of Florida campus, but also to dismiss his lawyers and defend himself; at the same time, allowing a number of other serial killers—Albert DeSalvo (the Boston Strangler) and David Berkowitz (the Son of Sam) to name just two—to be confined to mental institutions, DeSalvo without a criminal trial

- Allowing patients in mental institutions to refuse medication or other treatment. (The theory is that despite their commitment to a mental institution, they have some level of competence to make their own treatment decisions. Many members of the psychiatric community consider this practice to be the most peculiar kind of legal logic.)

For years, psychiatrists have been diagnosing defendants as mentally ill by the definitions of the *Diagnostic and Statistical Manual of Mental Disorders,* yet these people still have been tried and convicted as criminals—one can be mentally ill, yet still be considered competent and legally sane. Victims' rights organizations and prosecutors have clamored against defendants who "hide behind the insanity defense." This is perhaps the greatest medicolegal Catch-22 of the U.S. criminal-justice system: One would have to be crazy to do the kind of brutal, sick things some people do, but if the person is crazy, he or she can't be tried; therefore, the person may be mentally ill but not legally insane.

The fight over who is evil and who is sick has the ring of the famous medieval test to determine who is a witch: A woman arrested for being a witch was thrown in the water; if she drowned, that proved her innocence; if she floated, that proved her guilt, and she was stoned to death.

Clearly, psychiatric expertise is held in low regard in U.S. courtrooms. The Supreme Court, in its 1983 decision in *Barefoot v. Estelle,* turned the notion of scientific and medical expertise on its head. The defendant, convicted of murdering a police officer, was sentenced to death by the jury in a separate sentencing hearing. In the sentencing hearing, the jury had to take into consideration the probability that the defendant would commit further violent crimes and would constitute a threat to society if he were ever freed.

The state called two psychiatrists who predicted that the defendant would pose such a threat, despite the fact that the American Psychiatric Association and many forensic psychiatrists believe that making predictions of future violence is an inherently unsound practice, and that psychiatrists have no better ability to make such predictions than anyone else. The Supreme Court upheld the sentence, ruling that if laypeople can make a judgment about future violence, so, too, can psychiatrists and, therefore, the process of relying on such expert opinion is valid.

This begs the question, though: Why call the expert? An expert is supposed to be able to render an informed, educated opinion that the lay juror cannot.

If the Supreme Court throws out the rationale for expert psychiatric testimony, why not throw out the rationale for all expert forensic scientific testimony? Then we could go back to the time when only eyewitness identification—or misidentification, as many researchers tell us is more often the case than not—was the primary source of testimony; the time when prejudice and fear were enough for a jury to convict; the time of trial by ordeal.

Science is supposed to free us of passion and prejudice. Science is supposed to seek the truth. As can easily be seen, however, when science conflicts with politics, passion, and prejudice, it can be twisted, manipulated, and made to serve the uses of the majority and of the powers that be.

In short, the misuse of science, the uneducated and untrained use of science, the sloppy use of science, or the politically motivated use of science in the name of justice can be detrimental to liberty. The game of justice in this country, and in the world, is played for keeps. Every day, new scientific findings make it ever more complex. We must be ever vigilant.

Author's Note

Over the eight years the first edition of *Beyond the Crime Lab* was available, it was pointed out to me a few times that the book was being used as a supplemental reading for forensic science classes, mostly at the high school level, because forensic science texts are written at the advanced college level.

Because of this, I have created a teacher's guide to the book, which gives a little extra background on some chapters, as well as some study questions you can use in your classes.

You can get the study guide by sending a letter to the following address, written on your school's letterhead, telling me the name of the course and the approximate number of students taught each year:

Jon Zonderman
Beyond the Crime Lab Teachers' Guide
535 Howellton Road
Orange CT 06477

Index